AQA GCSE additional science

for AQA

Authors

Michael Brimicombe

Simon Broadley

Philippa Gardom-Hulme

Mark Matthews

Contents

How to use this book

Welcome to your AQA GCSE Additional Science revision guide. This book has been specially written by experienced teachers and examiners to match the 2011 specification.

On this page you can see the types of feature you will find in this book. Everything in the book is designed to provide you with the support you need to help you prepare for your examinations and achieve your best.

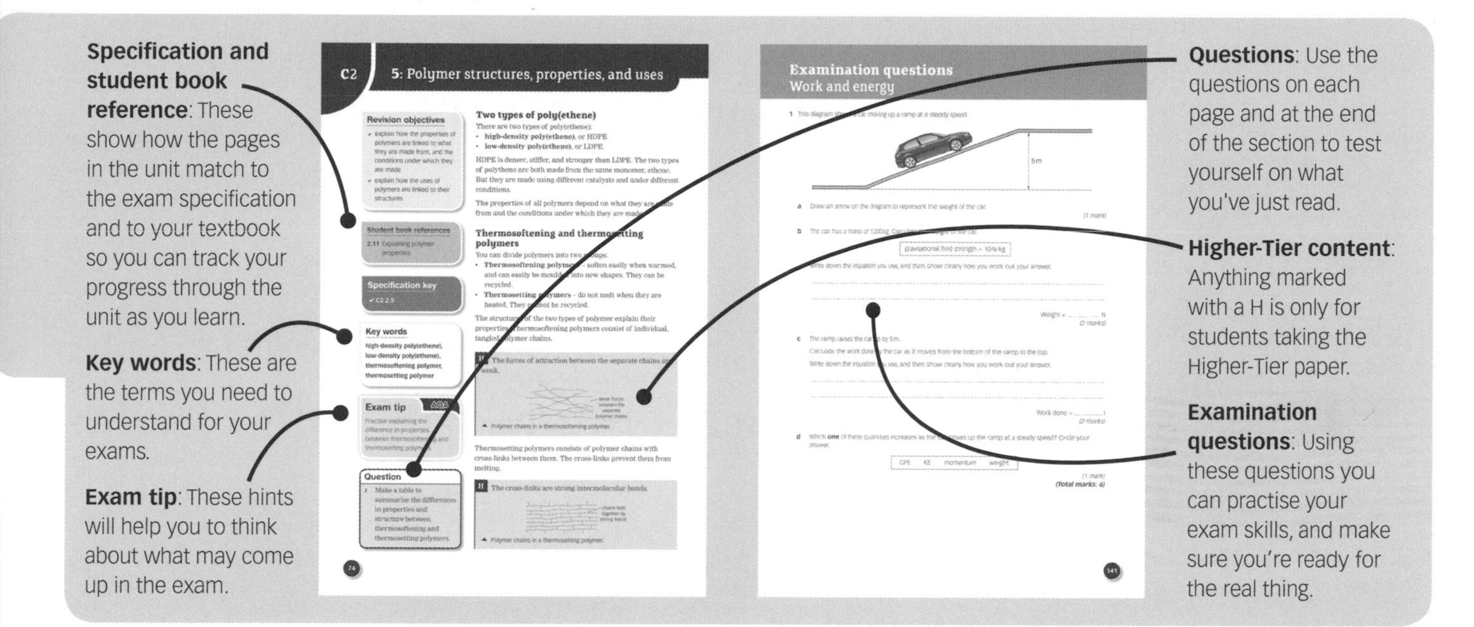

Specification and student book reference: These show how the pages in the unit match to the exam specification and to your textbook so you can track your progress through the unit as you learn.

Key words: These are the terms you need to understand for your exams.

Exam tip: These hints will help you to think about what may come up in the exam.

Questions: Use the questions on each page and at the end of the section to test yourself on what you've just read.

Higher-Tier content: Anything marked with a H is only for students taking the Higher-Tier paper.

Examination questions: Using these questions you can practise your exam skills, and make sure you're ready for the real thing.

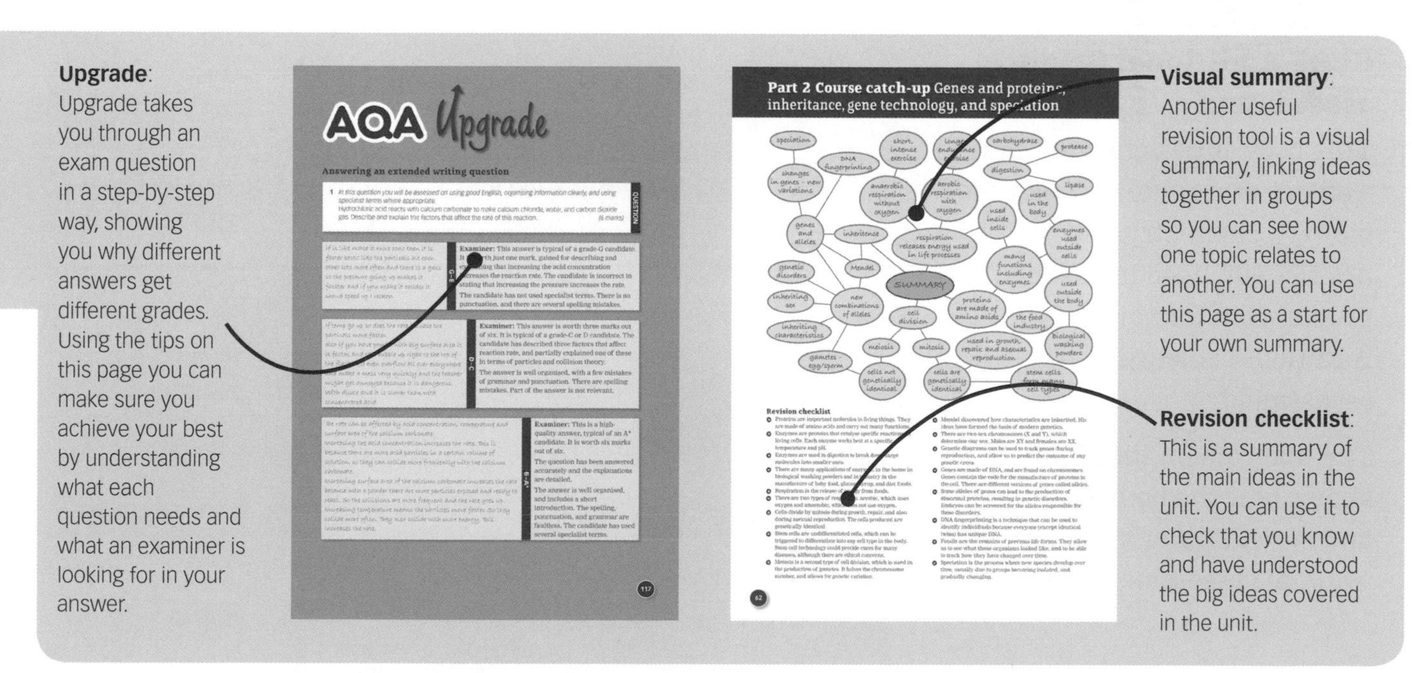

Upgrade: Upgrade takes you through an exam question in a step-by-step way, showing you why different answers get different grades. Using the tips on this page you can make sure you achieve your best by understanding what each question needs and what an examiner is looking for in your answer.

Visual summary: Another useful revision tool is a visual summary, linking ideas together in groups so you can see how one topic relates to another. You can use this page as a start for your own summary.

Revision checklist: This is a summary of the main ideas in the unit. You can use it to check that you know and have understood the big ideas covered in the unit.

Routes and assessment

Matching your course

The units in this book have been written to match the specification, no matter what you plan to study after your GCSE Additional Science course.

In the diagram below you can see that the units and part units can be used to study either for **GCSE Additional Science**, or as part of **GCSE Biology**, **GCSE Chemistry**, and **GCSE Physics** courses.

	GCSE Biology	GCSE Chemistry	GCSE Physics
GCSE Science	B1 (Part 1)	C1 (Part 1)	P1 (Part 1)
	B1 (Part 2)	C1 (Part 2)	P1 (Part 2)
GCSE Additional Science	B2 (Part 1)	C2 (Part 1)	P2 (Part 1)
	B2 (Part 2)	C2 (Part 2)	P2 (Part 2)
	B3 (Part 1)	C3 (Part 1)	P3 (Part 1)
	B3 (Part 2)	C3 (Part 2)	P3 (Part 2)

GCSE Additional Science assessment

This book is designed to offer a structured route through the biology, chemistry and physics you need to cover in preparation for your GCSE Additional Science assessment.

The diagram on the right shows how the book sections and spreads match the two available assessment routes.

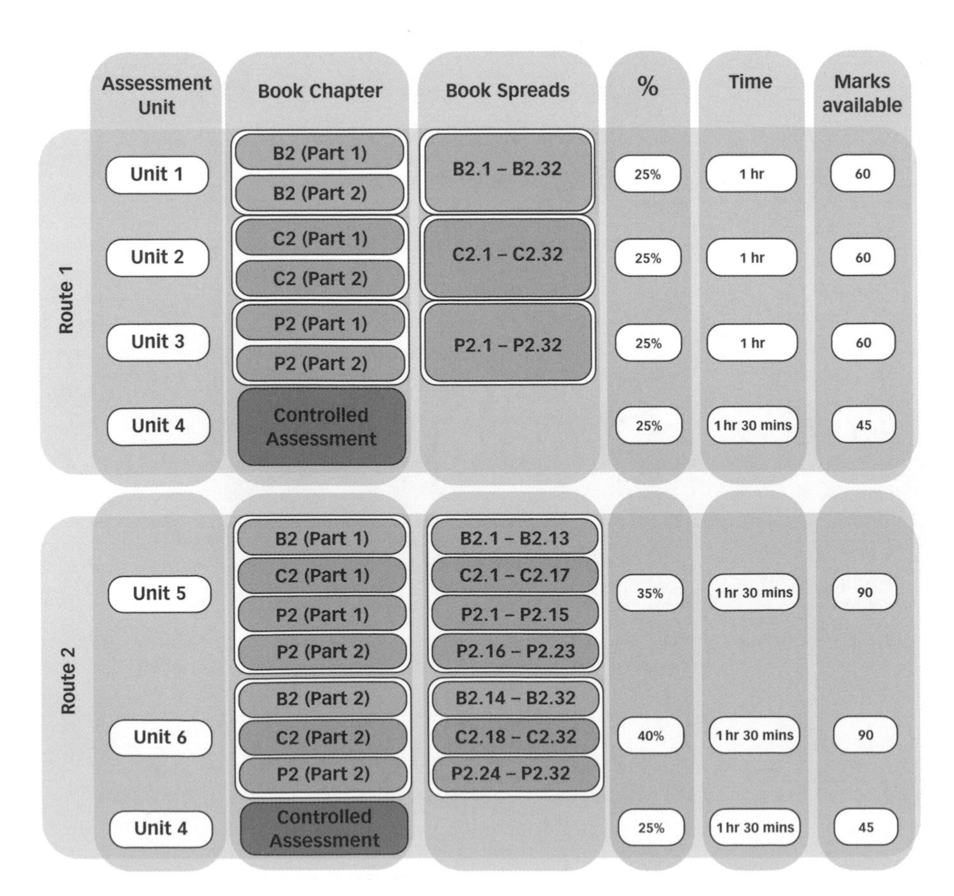

Understanding exam questions

When you read the questions in your exam papers you should make sure you know what kind of answer you are being asked for. The list below explains some of the common words you will see used in exam questions. Make sure you know what each word means. Always read the question thoroughly, even if you recognise the word used.

Calculate
Work out your answer by using a calculation. You can use your calculator to help you. You may need to use an equation; check whether one has been provided for you in the paper. The question will say if your working must be shown.

Describe
Write a detailed answer that covers what happens, when it happens, and where it happens. The question will let you know how much of the topic to cover. Talk about facts and characteristics. (Hint: don't confuse with 'Explain')

Explain
You will be asked how or why something happens.Write a detailed answer that covers how and why a thing happens. Talk about mechanisms and reasons. (Hint: don't confuse with 'Describe')

Evaluate
You will be given some facts, data or other information. Write about the data or facts and provide your own conclusion or opinion on them.

Outline
Give only the key facts of the topic. You may need to set out the steps of a procedure or process – make sure you write down the steps in the correct order.

Show
Write down the details, steps or calculations needed to prove an answer that you have been given.

Suggest
Think about what you've learnt in your science lessons and apply it to a new situation or a context. You may not know the answer. Use what you have learnt to suggest sensible answers to the question.

Write down
Give a short answer, without a supporting argument.

Top tips

Always read exam questions carefully, even if you recognise the word used. Look at the information in the question and the number of answer lines to see how much detail the examiner is looking for.

You can use bullet points or a diagram if it helps your answer.

If a number needs units you should include them, unless the units are already given on the answer line.

B2 1: Cells and cell structure

Revision objectives

- understand that living things are built from cells
- know the function of the parts of cells
- know the differences between different types of cells
- know that cells can become specialised for a specific function

Student book references

2.1 Plant and animal cells
2.2 Bacterial and fungal cells
2.3 Specialised cells

Specification key

- B2.1.1

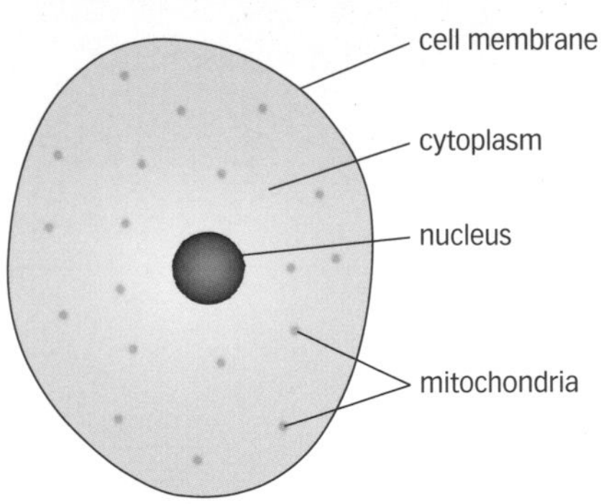

▲ A typical animal cell.

Exam tip

In this topic the key point is to know the parts of the cell, and what they do. Students often muddle them. Practise labelling cell parts on diagrams of the different types of cell, and give a function for each part.

The cell

The **cell** is the basic building block of all living things. Cells are so small that we need a **microscope** to see them.

Animal cell

Name of cell part	Function
Cell membrane	This is a thin layer around the cell. It controls the movement of substances into and out of the cell.
Nucleus	This is a large structure inside the cell. It contains chromosomes, which control the activities of the cell, and how it develops.
Cytoplasm	This is a jelly-like substance containing many chemicals. Most of the chemical reactions of the cell occur here.
Mitochondria	These are small rod-shaped structures that release energy from sugar during aerobic respiration.
Ribosomes	These are small ball-shaped structures in the cytoplasm, where proteins are made.

Plant cell

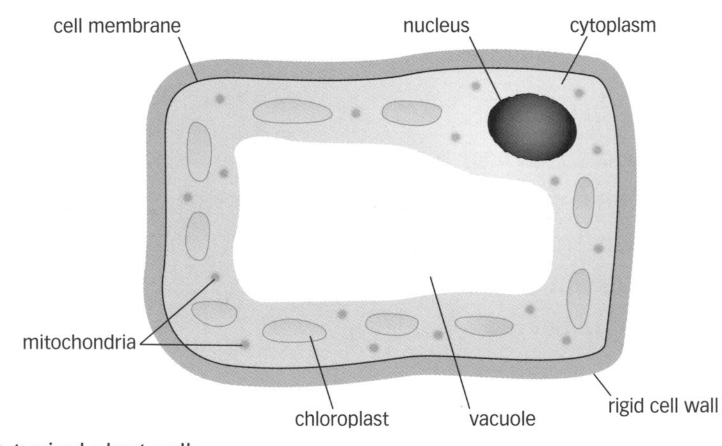

▲ A typical plant cell.

Plants have all of the structures of animal cells, plus a few others.

Part of cell	Function
Cell wall	This is outside the cell membrane. It is made of cellulose, which is strong and supports the cell.
Permanent vacuole	This is a fluid-filled cavity. The liquid inside is called cell sap, used for support.
Chloroplasts	These are small discs found in the cytoplasm. They contain the green pigment chlorophyll. Chlorophyll traps light energy for photosynthesis.

Other types of cell

Other groups of organisms are also made of cells, although the cells might have slight differences from plant or animal cells.

Algal cells

Algae are an important group of organisms, which includes the seaweeds. Their cell structure is the same as plant cells.

Bacterial cells

In **bacterial** cells, many of the structures found are similar to those of plants or animals, like the cytoplasm, membranes, and ribosomes. But:

- the cell wall has a similar function to a plant's, but it is made of a different chemical
- there is no distinct nucleus but the cell does have DNA, which is in the form of a loop.

Fungal cells

Again, the structures of the fungal cell are similar to those of plants and animals. They have a nucleus, cytoplasm, membranes, and cell wall.

Fungi are larger than bacteria and include important examples like the single-celled yeasts used in bread- and beer-making.

Special cells for special jobs

Although cells in a human or a plant have the same basic structures, they often carry out different jobs or functions. Cells become **specialised** to carry out a particular job, by developing special structures. This is called **differentiation**.

For example:

- Red blood cells – these lack most cell structures but contain large amounts of haemoglobin, which allows them to carry oxygen.
- Muscle cells – these contain contractile proteins, which allow the cell to shorten.
- Palisade cells – these contain many chloroplasts, which allow the cell to photosynthesise.
- Root hair cells – these have a long extension, which projects into the soil and increases the surface area for absorbing water.

Key words

cell, microscope, cell membrane, nucleus, cytoplasm, cell wall, permanent vacuole, chloroplasts, bacterial, fungi, specialised, differentiation

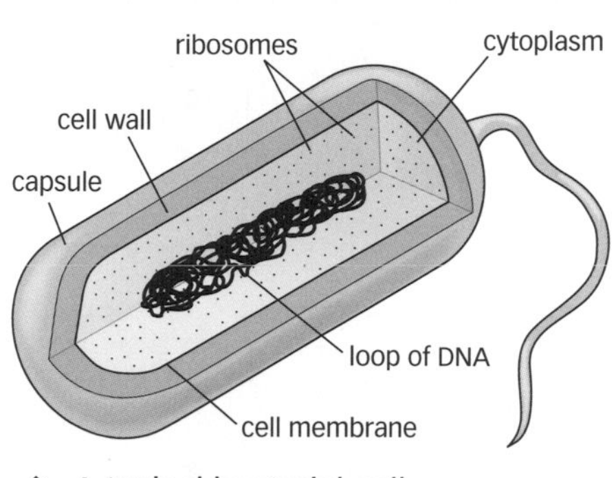

▲ A typical bacterial cell.

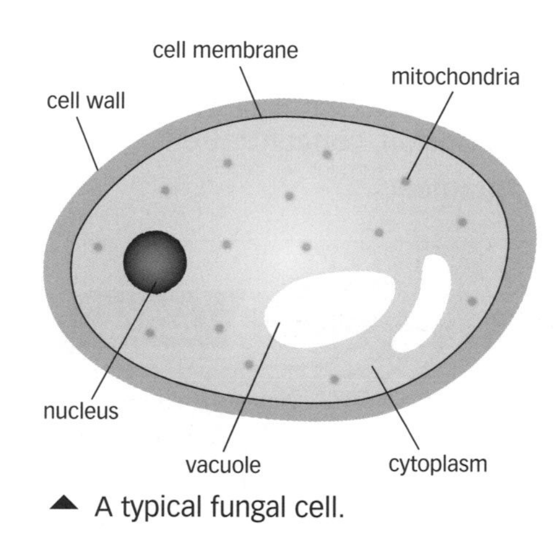

▲ A typical fungal cell.

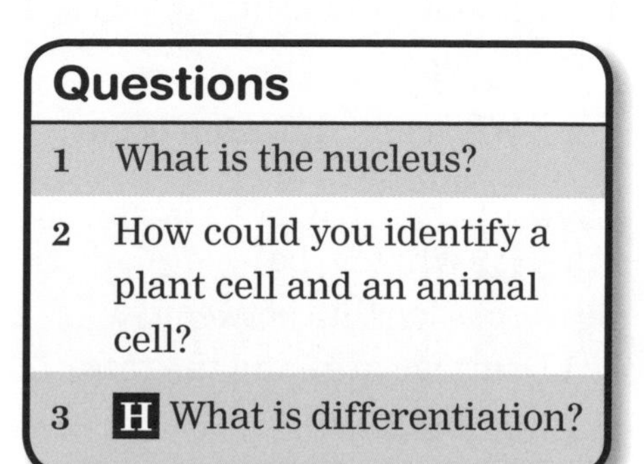

Questions

1 What is the nucleus?

2 How could you identify a plant cell and an animal cell?

3 **H** What is differentiation?

2: Diffusion

Revision objectives

- understand the process of diffusion
- know how diffusion allows particles to enter and leave cells

Student book references

2.4 Diffusion

Specification key

- B2.1.2

Key words

diffusion, concentration gradient

Exam tip AQA

Students don't usually find the idea of diffusion difficult, but they tend to make three mistakes. Remember: **1** the movement of particles is random; **2** it is the net movement; **3** the movement is from high concentration to low.

Questions

1 Why is diffusion important for living things?

2 Why can't diffusion occur in solids?

3 **H** What is the relationship between surface area and the rate of diffusion?

Movement of molecules

It is important for molecules to be able to move into and out of cells for them to work. **Diffusion** is one important method for achieving this.

Diffusion

Diffusion is the net movement of particles from an area of high concentration to an area of low concentration, until the concentration evens out. This happens in a liquid or gas where the particles can move.

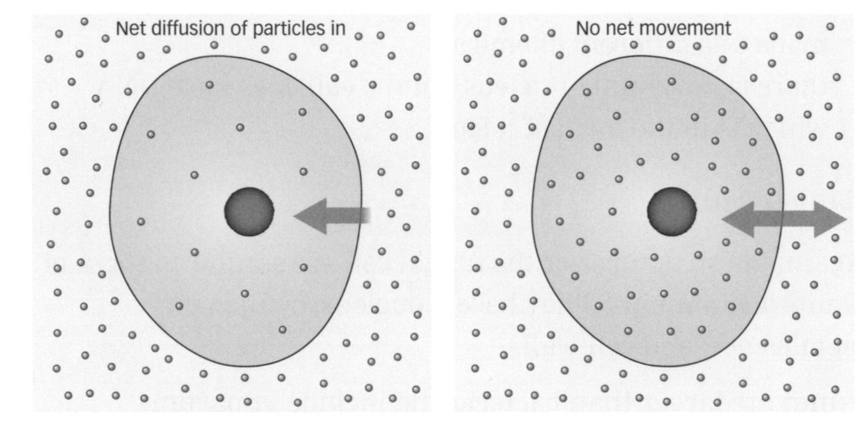

▲ Particles moving into cells by diffusion.

Examples of diffusion are:

- Oxygen diffuses into cells for use in respiration.
- Carbon dioxide diffuses out of cells as the waste product of respiration.

Factors affecting the rate of diffusion

Particles are constantly moving. Diffusion is the net movement of particles in one direction. Several factors affect the rate at which particles move:

- Distance – the shorter the distance the particles have to move, the quicker the rate of diffusion. For example, leaves are thin so carbon dioxide can move through the leaf quickly.
- **Concentration gradient** – particles move down a concentration gradient from high to low concentration. The greater the difference in concentration, the faster the rate of diffusion.
- Surface area – the greater the surface area means that there is more surface over which the molecules can move, so the rate is faster. For example, the lungs have a large surface area for the movement of oxygen.

Working to Grade E

1 Look at this diagram of an animal cell.

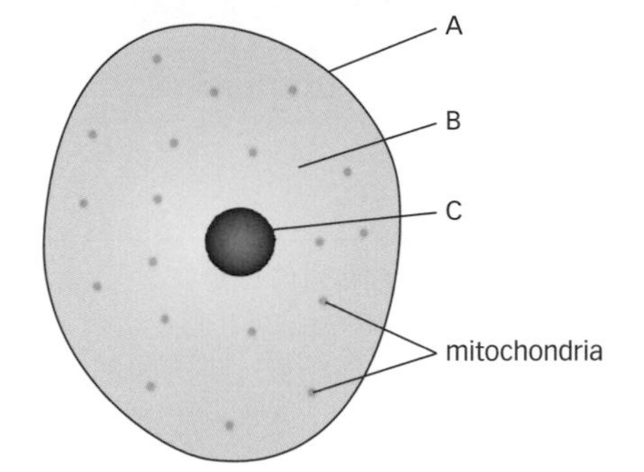

a Which of the labels A to C are:
  i the cell membrane
  ii the nucleus
  iii the cytoplasm?

b List **three** structures you would expect to find in a plant cell that are not present in the animal cell.

2 Where is the cell sap found in a plant cell?

Working to Grade C

3 What is the function of:
  a the nucleus
  b the cell wall
  c the cell membrane
  d the chloroplast?

4 Look at these drawings of three cells found in either plants or animals.

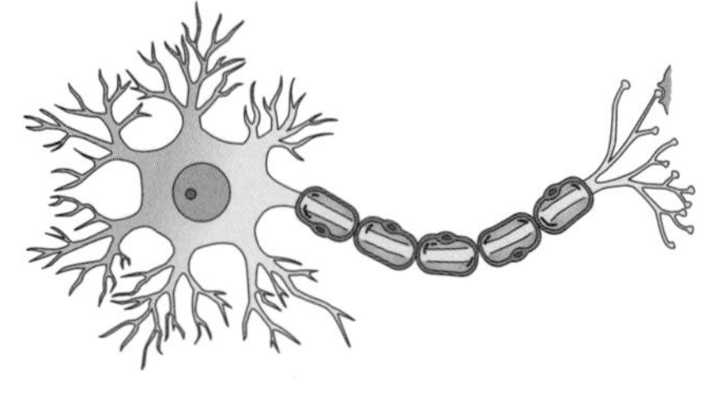

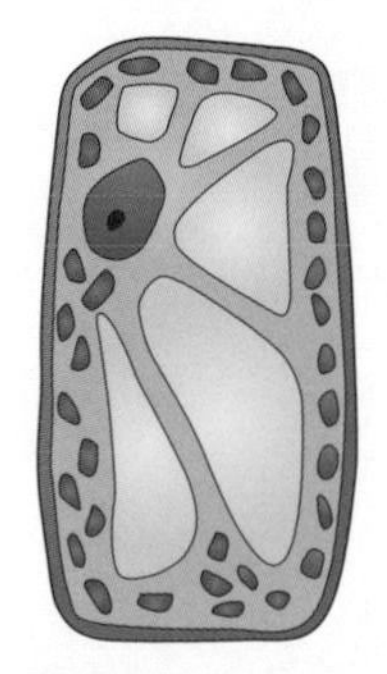

For each of the cells:
  a identify a special part of the cell
  b explain how this part helps the cell perform its function.

5 What is the name of the process where cells become specialised?

6 Draw and label a yeast cell.

7 Why are yeasts unable to photosynthesise?

8 Define diffusion.

9 Give **one** example of a molecule that moves into a cell by diffusion.

10 List **three** factors that might affect the rate of diffusion.

Working to Grade A*

11 Ribosomes are found in all cell types. What do they do?

12 Describe the differences between a bacterial cell and a plant cell.

13 Look at this diagram of a cell. There are different chemicals at each corner labelled A to D.

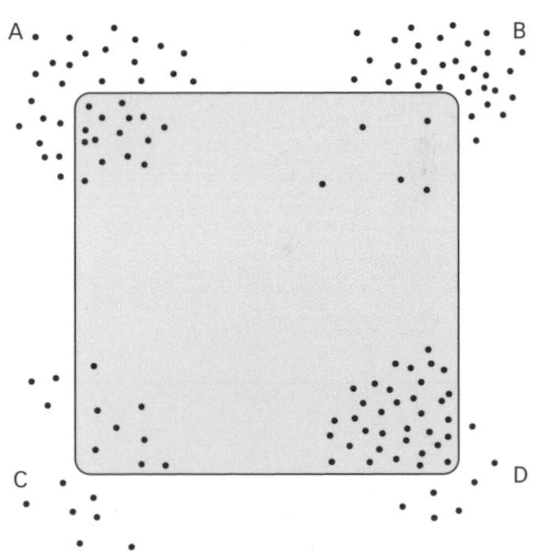

Indicate on the diagram whether diffusion is occurring at each corner, and if so in which direction it will occur.

14 Explain what is meant by a concentration gradient.

15 Diffusion occurs in cells.
  a Explain what is meant by the rate of diffusion.
  b Explain how the three factors from question 10, above, would affect the rate of diffusion.

16 Diffusion occurs efficiently in the lungs. Explain why this is so.

Examination questions
Cells and diffusion

1 Look at the table.

a Complete the table by filling in the name of either the structure or the function.

Structure	Function
Cell membrane	
	Controls the cell, contains DNA.
Cell wall	
	Releases energy in aerobic respiration.
Chloroplasts	
Vacuole	
	Where many chemical reactions occur.
	Proteins are made here.

(8 marks)

b Of the structures listed above, identify **three** that are only present in plant cells.

1 ....

2 ....

3 ....

(3 marks)

(Total marks: 11)

2 Diffusion occurs in cells.

a Define diffusion.

....

....

....

(2 marks)

b Name a gas that would diffuse into a cell for use in respiration.

....

(1 mark)

(Total marks: 3)

Working together

Animals and plants are **multicellular**, which means built of many **cells**. Cells do not work in isolation. The cells in our body are organised.

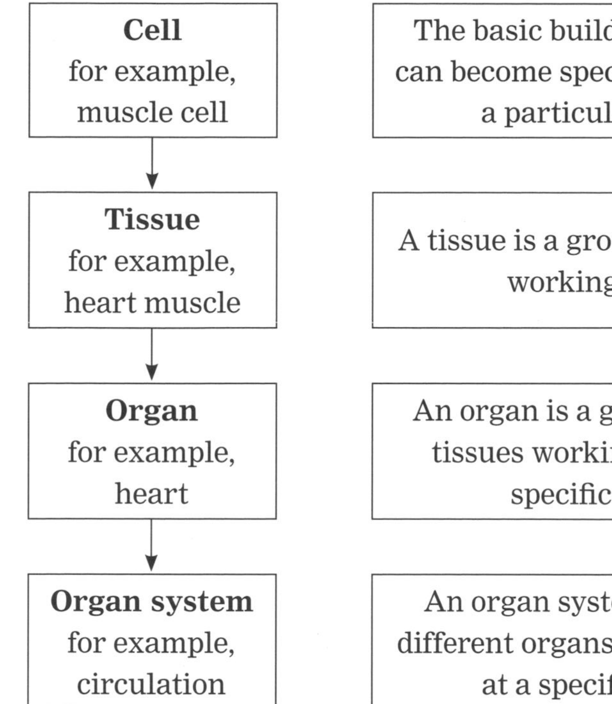

Cell for example, muscle cell	The basic building block, which can become specialised to perform a particular function.
Tissue for example, heart muscle	A tissue is a group of similar cells working together.
Organ for example, heart	An organ is a group of different tissues working together at a specific function.
Organ system for example, circulation	An organ system is a group of different organs working together at a specific function.

Revision objectives

- understand the levels of organisation of cells, tissues, and organs
- appreciate the working of some animal tissues and organs
- know that organs work together in organ systems
- know the major organs of the digestive system

Student book references

2.5 Animal tissues and organs

2.6 Animal organ systems

Specification key

- B2.2.1

Animal tissues

There are many examples of tissues in animals.

Muscle tissues – these are able to contract to bring about movement.
Glandular tissues – these produce substances like enzymes and hormones.
Epithelial tissues – these tissues act as covering for parts of the body.

Animal organs

A good example of where tissues work together in an organ is the stomach.

Epithelial tissue covers the outside of the stomach.

Muscular tissue, which can contract causing the wall of the stomach to move. This will churn the contents up.

Glandular tissue, which produces acid and enzymes that are poured into the stomach cavity to help digest food.

Inner epithelial tissue to cover the inside of the stomach.

▲ The stomach is an organ that has several different types of tissue working together.

Key words

multicellular, cell, tissue, organ, organ system, digestive system

Animal organ systems

Organs work together to form organ systems. An example of an organ system in the body is the **digestive system**. This is an exchange system that has two major functions in the body:

- Digestion – where food is broken down. Juices, containing enzymes, produced in glands are released into the gut to digest the food.
- Absorption – where useful molecules are taken from the gut into the blood.

The different organs include:

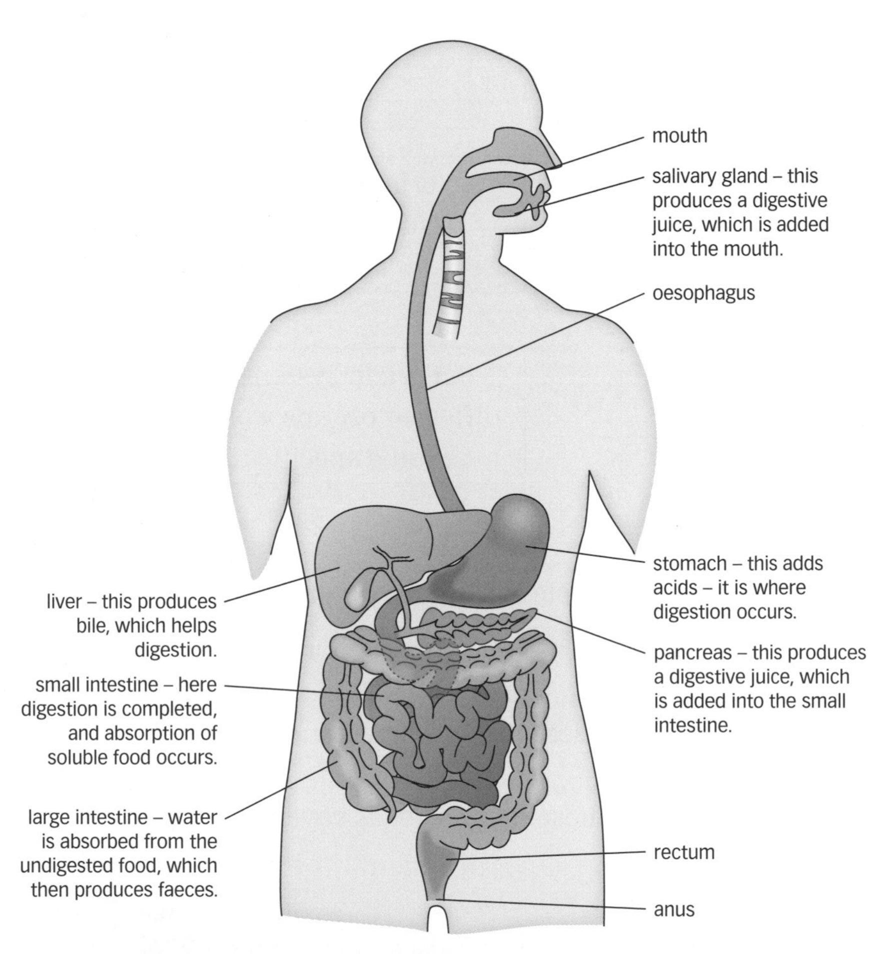

▲ In the digestive system several organs work together to bring about digestion.

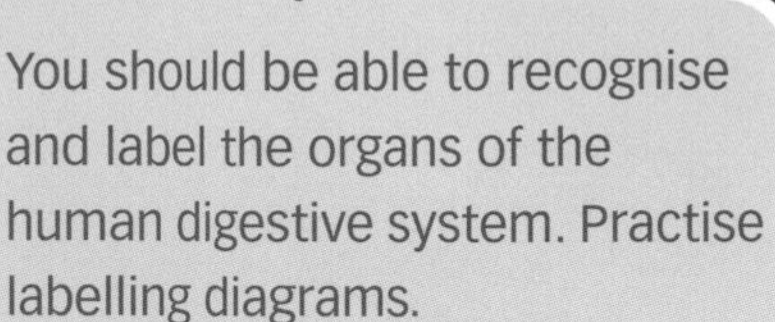

Exam tip AQA

You should be able to recognise and label the organs of the human digestive system. Practise labelling diagrams.

Questions

1. What does multicellular mean?
2. Name **three** organs in the human digestive system.
3. **H** What is the difference between an organ and a tissue?

Plant organs

Plants are also organised into tissues, organs, and organ systems.

Plant organs include:

Organ	Function
Stem	Supports the plant. Transports molecules through the plant.
Leaf	Produces food by photosynthesis.
Root	Anchors the plant. Takes up water and minerals from the soil.

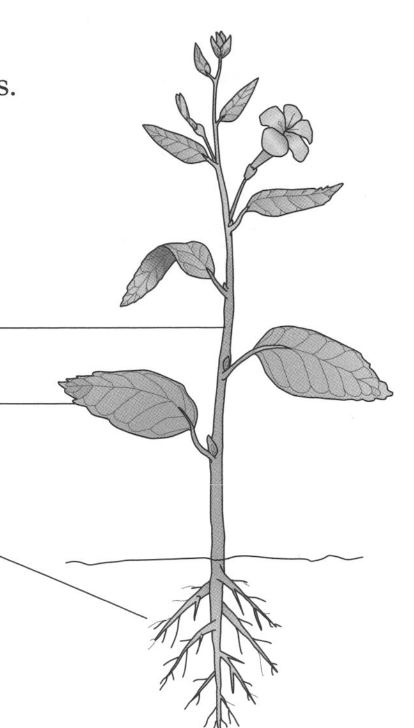

Revision objectives

- know that plant cells are organised into tissues
- know the main organs of the plant
- understand the distribution and functions of some of the key tissues of the plant

Student book references

2.7 Plant tissues and organs

Specification key

✔ B2.2.2

Key words

stem, leaf, root, epidermal tissue, mesophyll, xylem, phloem

Plant tissues

The organs of the plant are made from tissues, just as in an animal. Examples of plant tissues include:

- **epidermal tissues**, which form a covering layer over the surface of the plant
- a **mesophyll** tissue layer inside the leaf, which contains cells loaded with chloroplasts. These cells carry out photosynthesis.
- **xylem** tissues, which are made of hollow cells with strong cell walls. The cells are stacked one above the other and form a long tube through the plant. Xylem tissue is found around the edge of the stem. These cells carry water from the roots to the leaves, and help support the plant.
- **phloem** tissues found close to the xylem. Again phloem form long tubes through the plant. These cells transport sugars from the leaves to other parts of the plant.

Questions

1 Name **three** tissues of a plant.

2 Which tissues are involved in transport in plants?

3 **H** How is xylem adapted to carry out the function of transporting water?

Exam tip AQA

You need to be able to label the organs of a plant. You do not need to know the internal structure of these organs, except the leaf, which is detailed with photosynthesis.

Questions
Tissues and organs

Working to Grade E

1 Define a tissue.
2 Give an example of:
  a a tissue
  b an organ system.
3 Look at the drawings of the three human organ systems below.

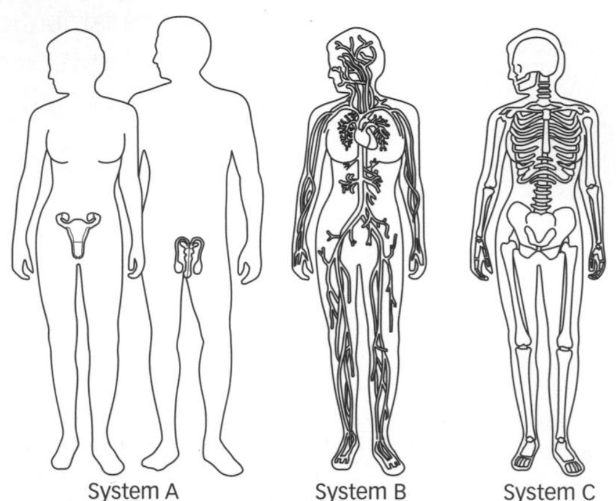

  a Name each organ system.
  b Name an organ found in system B.
4 Define an organ.
5 Digestion occurs in the digestive system. Where do the following events occur in the digestive system?
  a Food enters the system.
  b Water is absorbed from undigested food.
  c Digestion is completed and absorption occurs.
6 Below is a diagram of the digestive system. Label parts A to H.

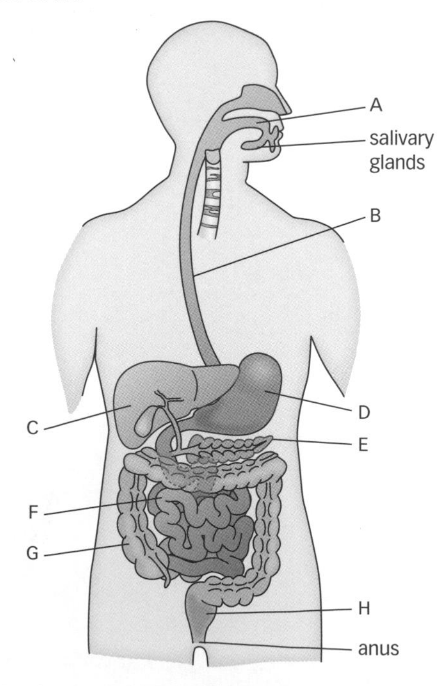

7 What is the function of the following plant organs:
  a stem
  b leaf
  c root

Working to Grade C

8 a What is the function of each organ system in question 3?
  b Why are they all called organ systems?
9 The stomach is an organ.
  a What is the role of the muscular tissue in the stomach?
  b Which type of tissue produces acid and enzymes?
  c Where would you find epithelial tissues in the stomach?
10 What happens in the following organs to help digestion?
  a the pancreas
  b the liver
  c the salivary glands.
11 Why is the flower regarded as an organ system?
12 What type of tissue covers the outside of roots, stems, and leaves?
13 What does the xylem tissue transport?
14 Draw a diagram of a stem to show where the xylem and phloem would be found.
15 How are the cells arranged in the xylem tissue?
16 Where is the palisade mesophyll tissue found?
17 What is the function of the palisade mesophyll tissue?
18 What substance is transported in the phloem tissue?

Working to Grade A*

19 Explain why three different types of tissue are required in the stomach.
20 Phloem is a tissue that transports sugars.
  a Where are these sugars picked up by the phloem?
  b Where does the phloem take the sugars to?
21 There are many tissues in plants.
  a Which tissue in the plant is involved in support?
  b How are the cells of this tissue adapted for support?

Examination questions
Tissues and organs

1 The stomach is an organ made of several tissues.

Below is a diagram of the stomach labelled with some of the tissues that make up the stomach.

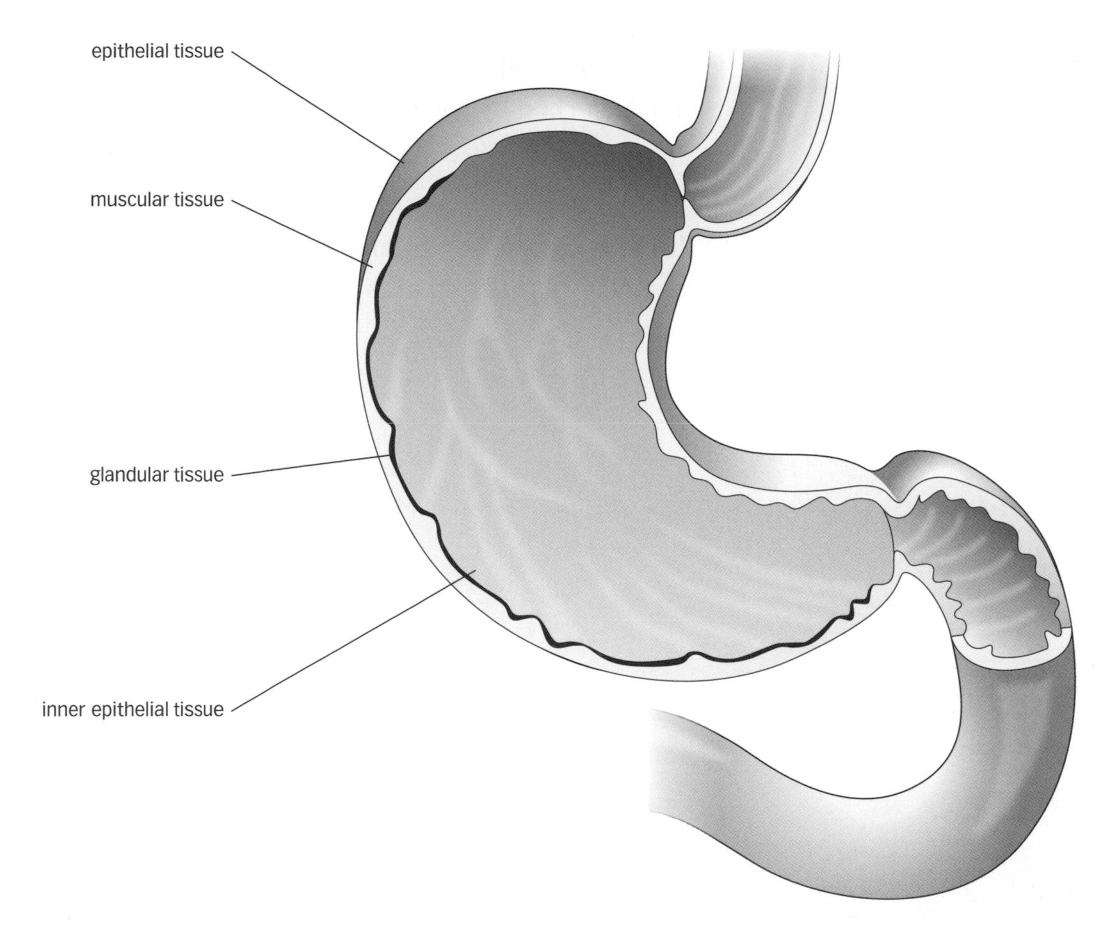

a Explain why three tissue types are needed for the correct functioning of the stomach.

....

....

....

....

....

....

....

....

....

....

(5 marks)

(Total marks: 5)

5: Photosynthesis

Revision objectives

- know that photosynthesis is the process by which plants make food
- know the sources of the raw materials for photosynthesis
- understand the fate of the products of photosynthesis
- appreciate the structure of the leaf as the site of photosynthesis

Student book references

2.7 Plant tissues and organs

2.8 Photosynthesis

2.9 The leaf and photosynthesis

Specification key

✔ B2.3.1 a – b, e – g

The process of photosynthesis

Photosynthesis is the process where plants make their food. They need two raw materials:

- carbon dioxide from the air
- water absorbed by the roots.

They also need:

- light energy from the Sun
- **chlorophyll**, a green pigment found in the chloroplasts in the cells of plants and algae, which absorbs the light.

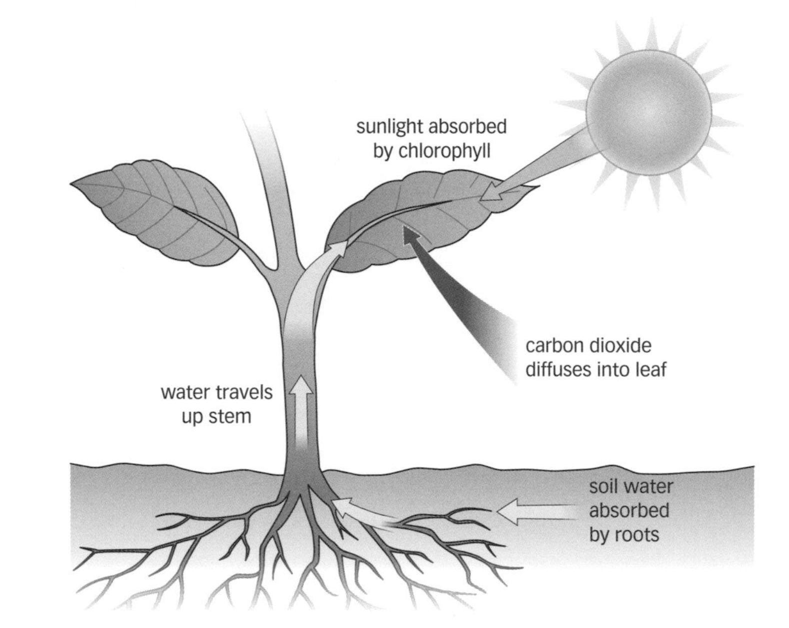

▲ In photosynthesis the plant uses sunlight energy to convert water and carbon dioxide into carbohydrates.

The equation for photosynthesis is:

$$\text{carbon dioxide} + \text{water} \xrightarrow[\text{chlorophyll}]{\text{sunlight}} \text{glucose} + \text{oxygen}$$

The light energy is needed to cause the reaction to happen between carbon dioxide and water, to make the carbohydrate glucose.

The products of photosynthesis

There are two products of photosynthesis:

- The main product of photosynthesis is the carbohydrate glucose.
- Oxygen is a by-product of the process and is released.

Glucose is the product that the plant needs. Some of it is used in respiration by the plant, but the rest is converted into other substances.

The conversion of glucose

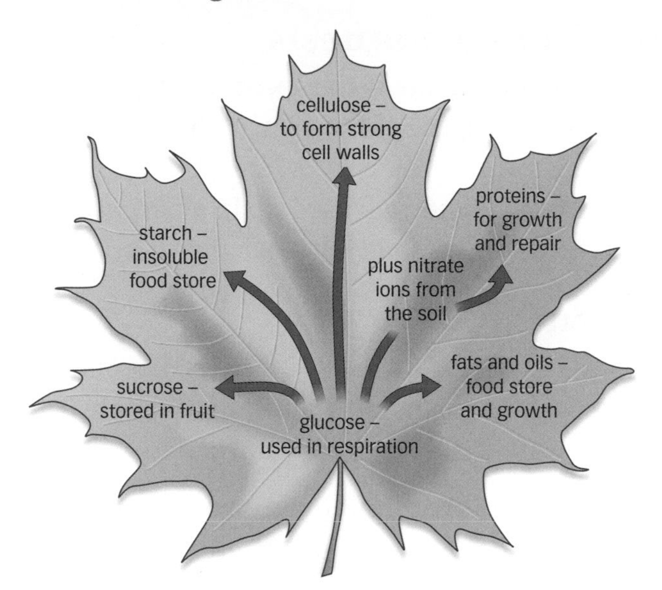

▲ Glucose from photosynthesis is converted into all the substances that a plant needs.

Key words

photosynthesis, chlorophyll, leaf, mesophyll, xylem, phloem, stomata

The leaf as the site of photosynthesis

The **leaf** is the main site of photosynthesis. It is well adapted for this function.

- The **mesophyll** contains the cells that carry out photosynthesis.
- The epidermis forms a covering layer.
- The **xylem** brings water to the mesophyll cells.
- The **phloem** takes the glucose away.
- There are **stomata** on the lower surface to allow carbon dioxide in and oxygen out.

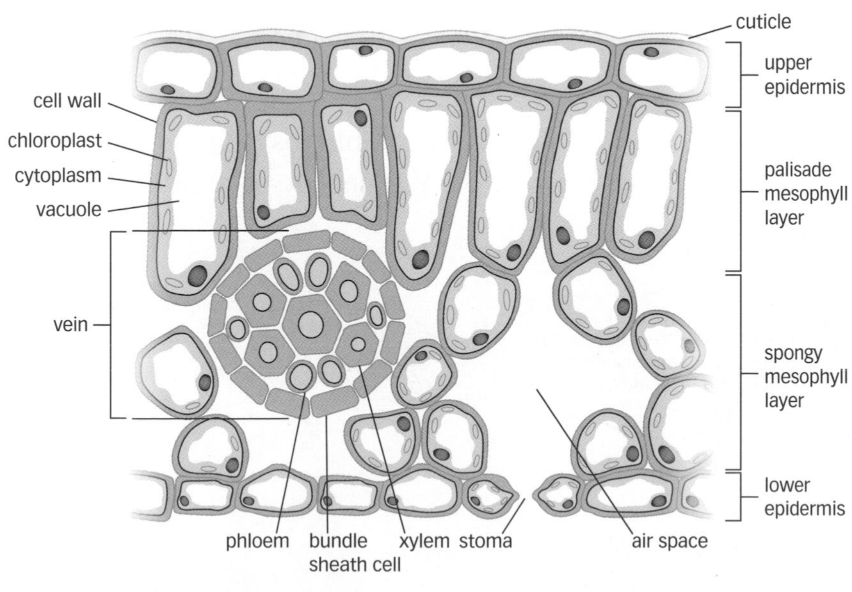

▲ The internal structure of a leaf.

Exam tip

AQA

Learn the equation for photosynthesis. Most of the questions on photosynthesis tend to ask about what is used, and what is made. You can answer these questions by knowing the equation.

Questions

1 What is the point of photosynthesis?

2 What is the energy change in photosynthesis?

3 **H** Describe how the two raw materials of photosynthesis get to the mesophyll cells in the leaf.

B2 6: Rates of photosynthesis

Revision objectives

- ✔ know how some factors affect the rate of photosynthesis
- ✔ understand limiting factors
- ✔ understand that some factors needed for photosynthesis can be controlled
- ✔ appreciate the commercial benefits of controlling photosynthesis in greenhouses

Student book references

2.10 Rates of photosynthesis

2.11 Controlling photosynthesis

Specification key

✔ **B2.3.1 c – d**

The rate of photosynthesis

The **rate of photosynthesis** is the speed at which a plant photosynthesises. Biologists can measure this in one of two ways:

- the amount of raw materials used up in a period of time
- the amount of product made in a period of time.

Limiting factors

When a process is affected by several factors, the one that is at the lowest level will be the factor that limits the rate of reaction. This is called the **limiting factor**.

There are three factors that limit the rate of photosynthesis:

- Availability of **light** – the less light there is, the slower the rate of photosynthesis.
- A suitable **temperature** – temperature affects the enzyme reactions. As the temperature increases so does the rate, but if the temperature becomes too high it will damage the enzymes and stop photosynthesis.
- The amount of **carbon dioxide** – the less carbon dioxide, the slower the rate of photosynthesis.

If the limiting factor is increased, then the rate of photosynthesis will increase, until one of the other factors becomes limiting.

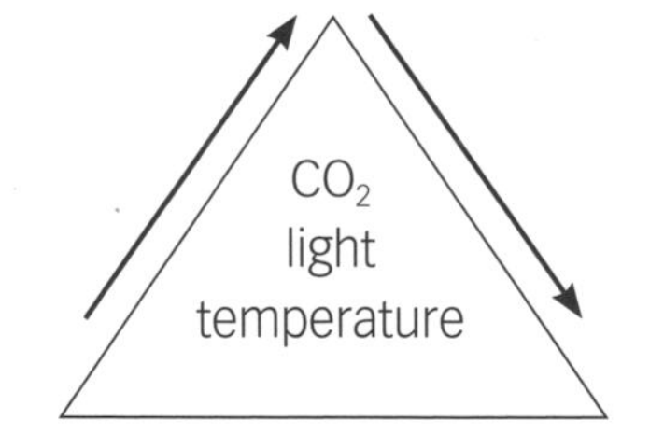

DECREASING the factor DECREASES the rate

One way of showing how the rate of photosynthesis is affected by a factor is by using a graph. You need to be able to interpret these graphs.

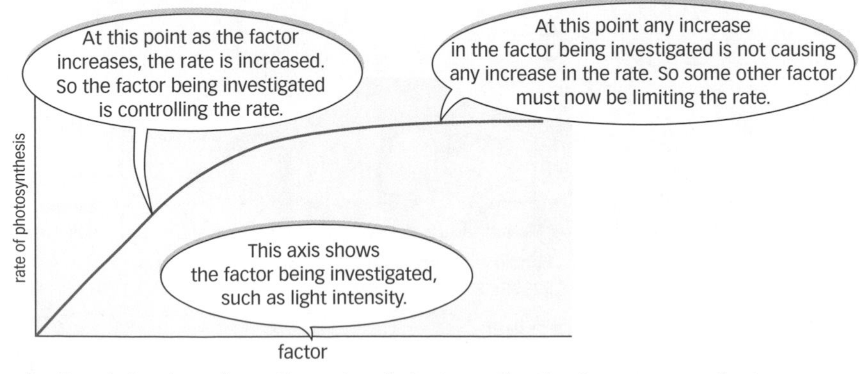

▲ Graph to show how the rate of photosynthesis changes as a factor increases.

Controlling photosynthesis

Gardeners and plant growers use **greenhouses** to be able to grow crops all year round, and to grow tropical plants. This is because we can control the environment inside the greenhouse to maximise the rate of photosynthesis.

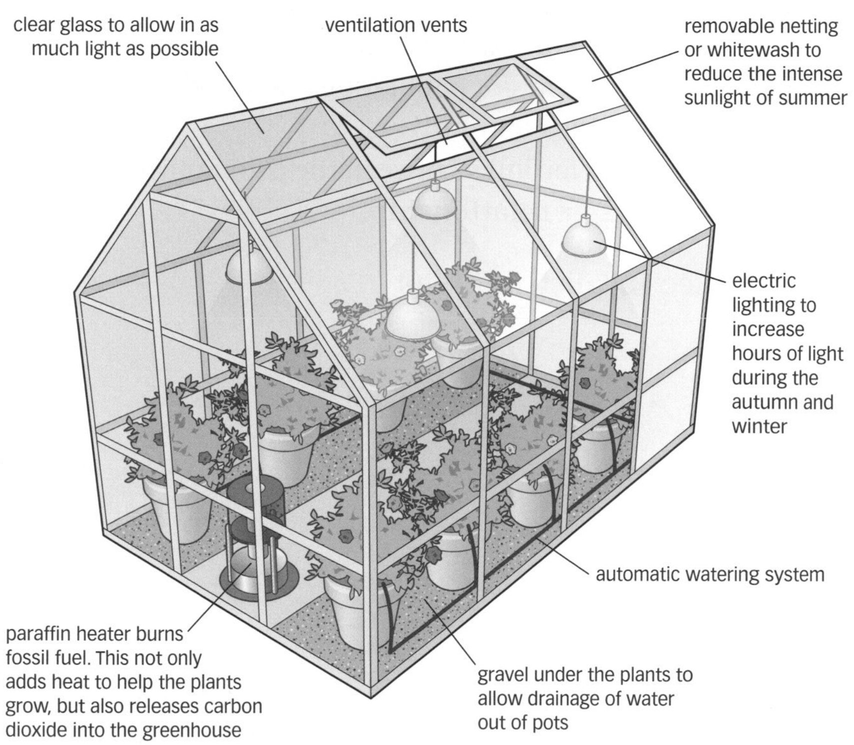

▲ Modern greenhouses use automated systems to give the best conditions for photosynthesis.

The gains and costs of greenhouses

Greenhouses are used to allow plants to be grown all year round.

Factor controlled	Advantages	Disadvantages
Light	Increasing light increases the rate of photosynthesis, increasing the yield, giving greater profits.	Cost of electric lighting. Cost of nets.
Temperature	Warmth increases the rate of photosynthesis, especially during cold months. So plants can be grown out of season when they have greater value.	Glass is expensive. Cost of fuel.
Carbon dioxide	Adding carbon dioxide from burning fuels speeds up photosynthesis, increasing yield.	Cost of fuels.
Water	The correct amount of water stops plants dying and lost income.	Cost of electricity to run automated systems.

Key words

rate of photosynthesis, limiting factor, light, temperature, carbon dioxide, greenhouse

Exam tip AQA

Remember to say that if the factor increases, then the rate will increase. Students often explain an increase in rate by just naming the factor and forget to say that the factor must increase.

Questions

1 Why is photosynthesis faster on a sunny day?

2 How do gardeners use greenhouses to alter the rate of photosynthesis?

3 **H** What changes should be made to the limiting factors to speed up plant growth?

7: Distribution of organisms

Revision objectives

- ✔ know about common sampling techniques
- ✔ understand how to use sampling techniques to collect good quality data
- ✔ know that environmental work generates large amounts of data
- ✔ understand some methods used to process data

Student book references

2.12 Sampling techniques
2.13 Handling environmental data

Specification key

✔ B2.4.1

▲ A transect line is a long tape that is laid through an environment. Quadrats can be placed at regular intervals along the tape; this prevents bias. A quadrat is a square frame. They can be placed at regular intervals along a transect line to obtain unbiased data, or placed randomly in an area. The numbers and types of organism are recorded in the quadrat. This supplies the quantitative data.

Living in the environment

Plants and animals live in many environments on Earth. Biologists need to make sense of where things live.

- They note where a species lives – this is the **distribution**.
- They count the number of individuals of a species in an area – this is the **population**.
- They look at the different populations that live together in an area – this is the **community**.
- Finally they look for links between the community and external factors – a **relationship**.

Factors affecting distributions

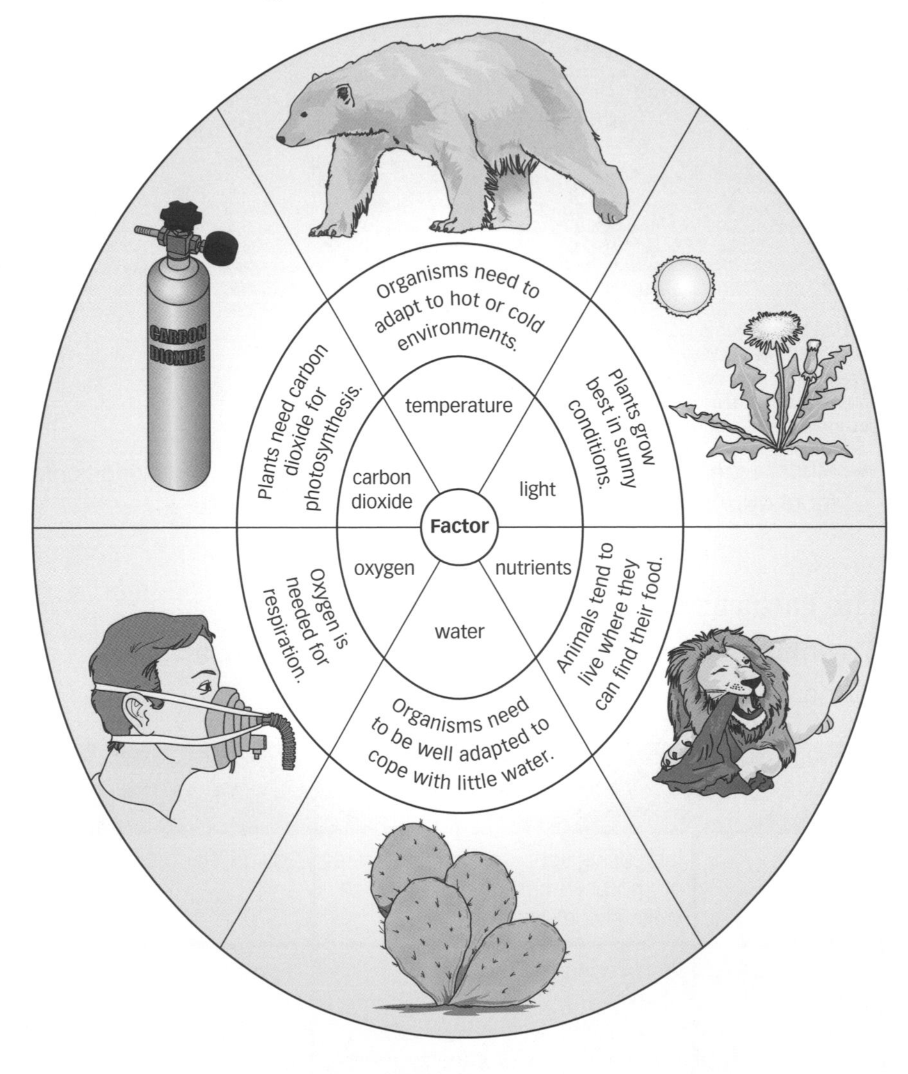

Sampling the environment

Biologists need to record data about the distribution of organisms. With this quantitative data they can begin to look for relationships concerning the cause of the distribution.

How can they record accurate unbiased data about an organism's distribution?

Obtaining valid data

Being accurate	Use appropriate apparatus for the task, as this will generate accurate results. Each recording should be a sufficiently large sample.
Being reliable	Take repeat readings. Repeats make results more reliable.
Being fair	Always use the same equipment for each test. Make sure that recordings are not biased. To do this, use regular points along a transect, or random **sampling**.

Key words

distribution, population, community, relationship, sampling, mean, median, mode

If the data is collected in this way, it should be both valid and repeatable by other workers. Only then can any conclusions be accepted.

Analysing the data

Data collection of this type generates lots of numbers. Biologists often analyse the data to make sense of it. This may mean looking for a central value to illustrate the data, for example, the **mean** size of a limpet's shell.

There are three types of central values:

- **Mean** – this is the average value.
- **Median** – this is the middle value of the data when ranked.
- **Mode** – this is the most common value.

Making sense of the data

An example of this is where biologists have tried to explain the changing distribution of a species in the UK, like the ringed plover.

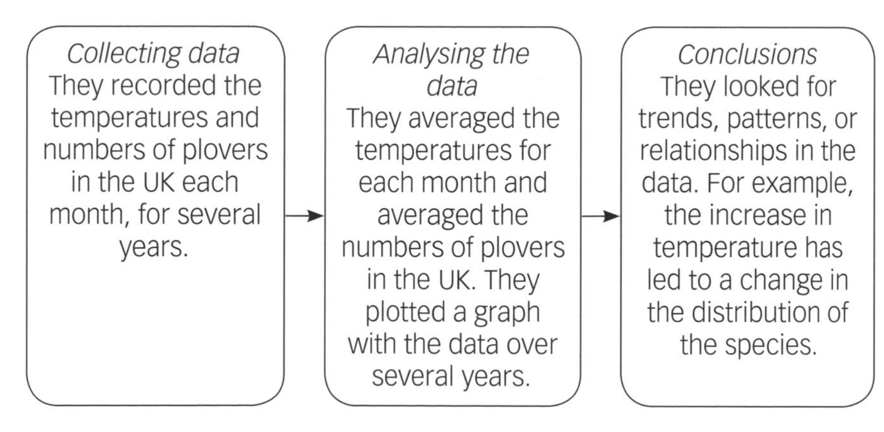

It is always important to make sure that the method used is valid, otherwise it can lead to incorrect conclusions being drawn.

Questions

1. How might you collect data about the distribution of organisms?
2. What three centralising values are used to analyse data?
3. What controls where an organism might live?

Exam tip AQA

When asked to explain the distribution of an organism, look for a relationship between where the organism is found and some trend in a physical factor.

Questions
Photosynthesis and distribution

Working to Grade E

1 What are the raw materials needed for photosynthesis?

2 What are the **two** products of photosynthesis?

3 What is the source of energy for the reactions of photosynthesis?

4 Name the pigment that traps the energy of photosynthesis.

5 Which organ of the plant is the site of most of the photosynthesis in the plant?

6 Name **two** groups of organisms that carry out photosynthesis.

7 Below is a drawing of a section through the leaf.

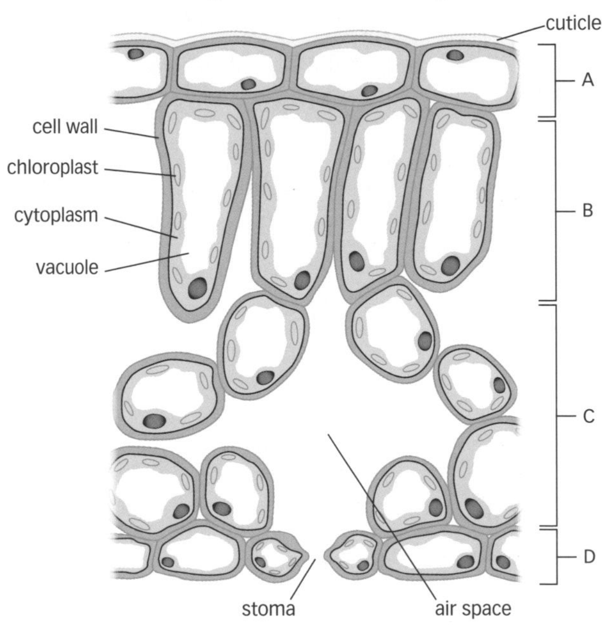

a Name the parts labelled A to D.
b Which tissue in the leaf is where most photosynthesis occurs?

8 What is a population?

9 Light is important to plants.
a How does light affect the distribution of plants?
b Explain why this is the case.

10 What is a quadrat used for?

11 What is the relationship between temperature and the distribution of polar bears?

12 What is a community?

Working to Grade C

13 Write out the word equation for photosynthesis.

14 Look back at question 7. Where does carbon dioxide enter the leaf?

15 What do we mean by the term 'limiting factor'?

16 Define the term 'rate of photosynthesis'.

17 Grass plants grow on a roadside verge.
a On a typical sunny day in June, what is likely to be the limiting factor for a plant?
b On a frosty day in November, is the same factor still likely to be the one limiting the plant?

18 Look at this drawing of a greenhouse.

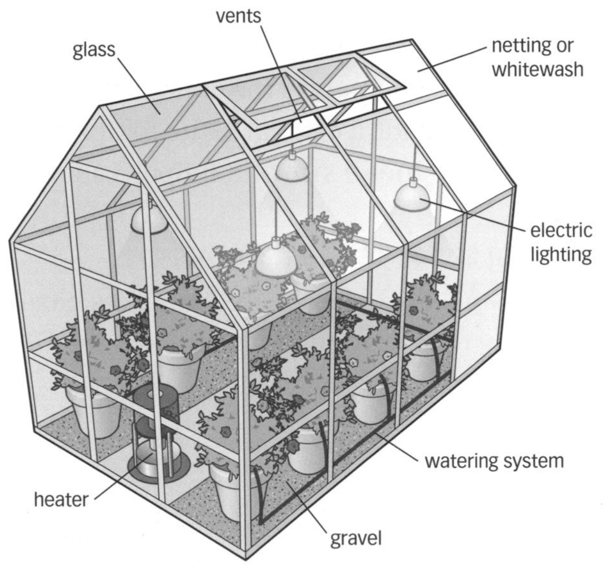

a A paraffin heater is shown in the greenhouse.
i What **two** things are supplied to the greenhouse by the use of a paraffin heater?
ii Explain why these products of the paraffin heater should be important to increase plant growth.

b Explain why it is important that the windows in the greenhouse can open.

c Why is there gravel under the pots in the greenhouse?

19 Why is electric lighting often not considered commercially viable in greenhouses?

20 Why do organisms have to be very highly adapted to live in a desert?

21 A group of students decided to study the distribution of daisies in a school playing field. They worked in three groups; each placed a quadrat at regular intervals starting at the school building and working out into the open field. They calculated the average of their data. The results are below.

Distance from school (m)	Average number of daisies
5	5.2
10	5.8
15	8.9
20	20.1
25	25.3
30	28.3
35	27.0

- **a** Why did the students take three sets of readings for each distance?
- **b** How did the students ensure that there was no bias in the data they collected?
- **c** What trend can be seen in the data?
- **d** Suggest a reason for this trend.

22 Bison are herbivores.

- **a** What is the relationship between the distribution of the bison and the numbers of grassland plants?
- **b** If the bison overgraze the result is a decrease in the numbers of grass plants. What will happen to the distribution of the bison?

23 What is the difference between a mode and a median?

24 Scientific evidence must be valid and repeatable.

- **a** How does sample size affect validity?
- **b** Why is it important that experimental results are repeatable?

Working to Grade A*

25 Look back at question 7.

- **a** Draw an arrow on the diagram to show the path taken by carbon dioxide to get to the photosynthesising cells.
- **b** Describe how the products of photosynthesis are removed from the leaf.

26 The products of photosynthesis can be converted into other molecules.

- **a** Proteins are one such molecule.
  - **i** What is added to the products of photosynthesis to make a protein?
  - **ii** Where does this additional substance come from?
- **b** What does the plant make to strengthen plant cell walls?
- **c** Why does the plant produce fats and oils?
- **d** What molecules make fruit sweet?

27 What substance does the plant use to carry out respiration?

28 Why is starch a more suitable storage compound than sugar?

29 The graph below shows how light intensity affects the rate of photosynthesis.

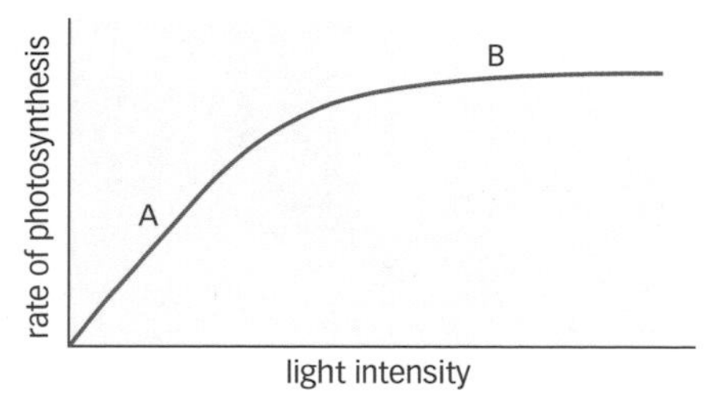

- **a** At which point on the graph, A or B, is light the limiting factor?
- **b** Explain why.
- **c** Why does the rate of photosynthesis not continue to increase?

30 Look at question 17, above.

- **a** When is the rate of photosynthesis going to be at its highest?
- **b** Explain your answer to question 17 part a.

31 Light is important to plants.

- **a** Explain how light is controlled in a greenhouse.
- **b** Explain why it is important for plant yield to control the light.

32 Explain how the use of a greenhouse to grow strawberries increases a farmer's profits.

33 Look at question 21, above. What other measurements would you need to take to prove your theory?

Examination questions
Photosynthesis and distribution

1 Use the words below to complete the diagram by filling in the empty boxes.

light	**water**	**Sun**	**carbon dioxide**	**glucose**	**oxygen**

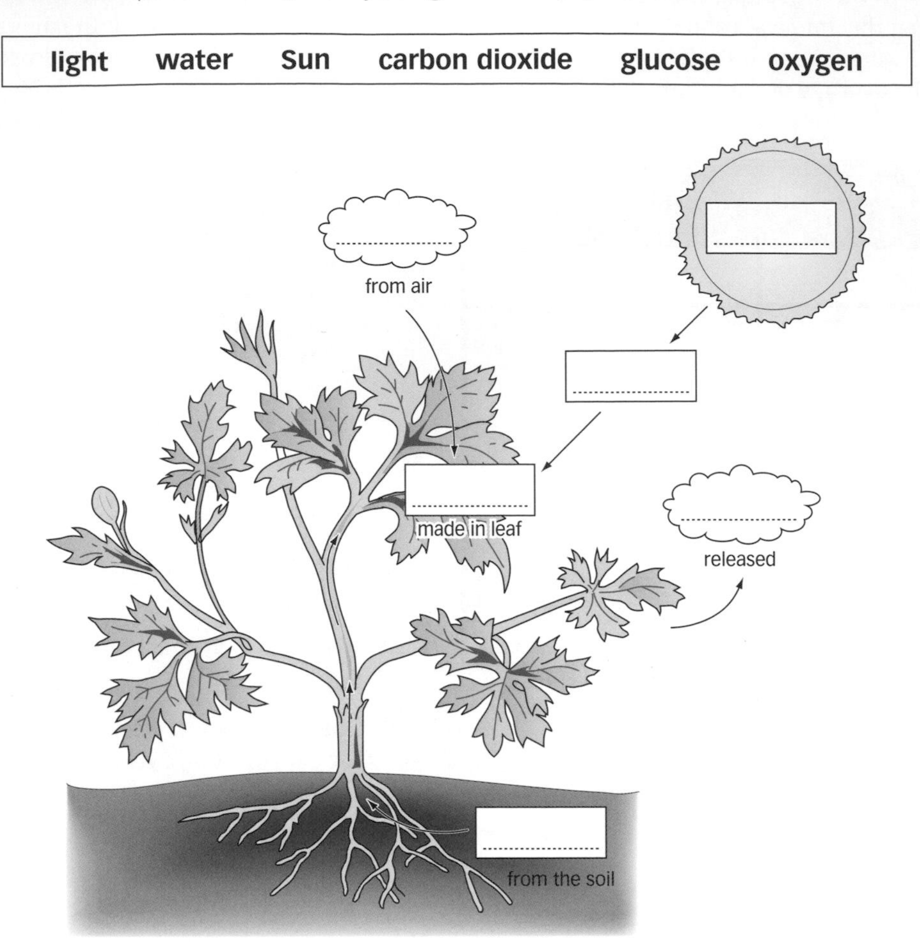

(6 marks)

(Total marks: 6)

2 A student investigated the effect of light intensity on the rate of photosynthesis. Below is a diagram of the apparatus that they used.

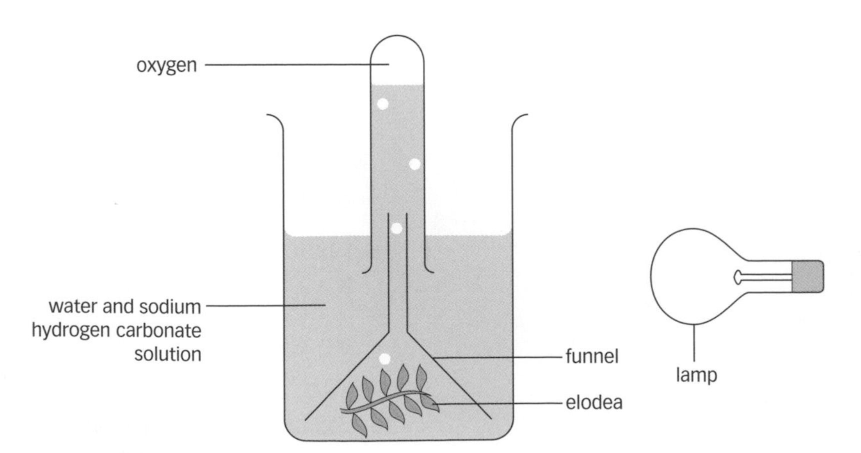

The student measured the light intensity, and the amount of oxygen produced as bubbles of oxygen produced per minute. The results are given below.

Light intensity (arbitrary units)	1	2	3	4	5	6	7
Number of bubbles of oxygen produced per minute	9	19		46	60	67	68

a Plot the data on the graph below, and join the points with a line.

(3 marks)

b Estimate a value for the reading at a light intensity of 3.

....

(1 mark)

c Explain fully the relationship between light intensity and the rate of photosynthesis.

....

....

....

....

....

(3 marks)

d What is a limiting factor?

....

....

....

(2 marks)

e Name **two** other limiting factors of photosynthesis.

1 ....

2 ....

(2 marks)

(Total marks: 11)

3 A student set up an experiment to study the gases given off by leaves. They used a bicarbonate indicator that was sensitive to the levels of carbon dioxide.

The indicator shows the following colour range:

Purple	Very low carbon dioxide levels
Red	Medium carbon dioxide levels
Yellow	Very high carbon dioxide levels

Below are the results of the experiment.

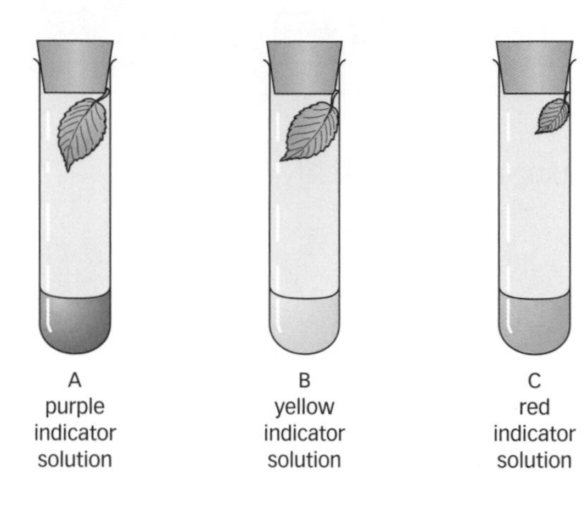

a Look at tubes A and B.

i Which one of these tubes was kept in the light?

....

(1 mark)

ii Explain your answer.

....

....

....

(3 marks)

b Why was it important that the two leaves in tubes A and B were the same size?

....

....

(1 mark)

c Why did the student avoid breathing heavily into the tubes when they set up the experiment?

....

....

....

(2 marks)

d Tube C was kept in the same conditions as tube A.

i Why was there a difference in the result?

....

....

(2 marks)

ii What might be the result in tube C if the leaf was killed by boiling before the experiment?

....

....

(1 mark)

(Total marks: 10)

4 A student wanted to conduct an experiment that would record the distribution of seaweed down a seashore.

a Describe a method the student could use to carry out this experiment. Identify any key equipment that might be needed.

....

....

....

....

....

....

....

....

....

(5 marks)

b How could the student ensure that the results would be:

i free from bias?

....

....

(1 mark)

ii reliable?

....

....

(1 mark)

(Total marks: 7)

Cells and the growing plant

Designing investigations to test a hypothesis

Scientists often start an investigation by suggesting a hypothesis. This might be a suggested relationship between two variables, in fact a prediction.

Whilst studying the ideas in this module, there are some good examples of investigations where you might be asked to produce a prediction that you could then test.

Look at the example that follows; this will show you how scientists design investigations to test their hypothesis. It will also highlight how scientists consider the validity of their data-collecting methods.

Distribution of Dog's mercury

Dog's mercury is a common woodland plant in Britain. It grows throughout the woodland, but seems to favour clearings rather more than ground under the tree's canopy.

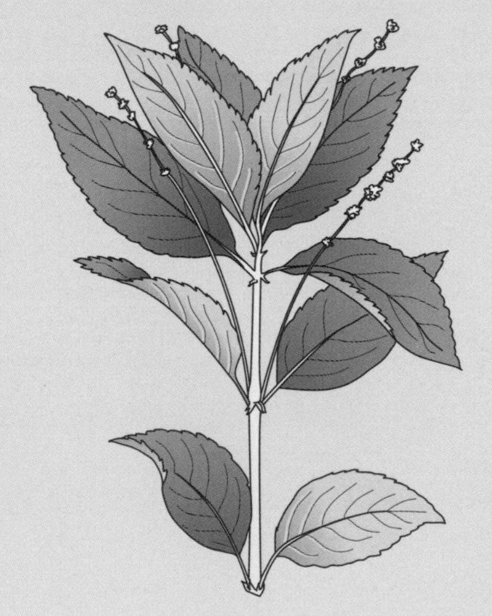

1 Propose a hypothesis for the distribution of Dog's mercury.

Skill – Producing a hypothesis

A hypothesis should have two parts:

A prediction – here it might be that Dog's mercury would grow better in higher light intensities.

- This identifies the two key variables: light and the growth of the plant.
- The prediction also gives a relationship between the two variables: more light = better plant growth.

A scientific explanation for the hypothesis – here it might be that where there is more light, there will be a higher rate of photosynthesis. This will make more food for the plant to grow.

- This explanation links a scientific process to the relationship given in the prediction.

2 In a woodland, your task is to design an investigation to test the hypothesis.

a What are the variables that you will measure?

Here you need to identify the variables:

- light intensity
- number of plants/percentage cover of Dog's mercury.

The two variables should come from your prediction.

b Given the following apparatus – a quadrat, a transect line, and a light meter – outline the method you would use to test the hypothesis.

Questions that ask you to design a method are looking for several clear and logical steps to be suggested. They should come from information on three key points:

- Do you know what each piece of apparatus is used for?
  - Place a transect line through the wood.
  - Place a quadrat at regular intervals along the transect line.
- Record quantitative data from the quadrats.
  - Take light readings in each quadrat.
  - Record the number of Dog's mercury plants or percentage cover of the plant in each quadrat.
- What will you keep the same?
  Include any two of the following:
  - size of the quadrat
  - distance between the quadrats
  - number of light readings in each quadrat
  - method for recording the plants in each quadrat.

c How would your method ensure that your results were valid in the following ways:

- avoiding bias?
- producing reliable, reproducible results?

To avoid bias you need to have a random system of taking your readings. This could be by placing quadrats randomly, or by placing them at regular pre-fixed points, such as at fixed distances from each other.

Reliable or reproducible results are produced by having sufficient repeats. (Remember 'For reliability do repeats'.) That will give you a large sample size in your results.

Answering a question where command words are important

QUESTION

In this country most tomatoes are grown commercially in greenhouses.

1. Name **three** conditions that can be controlled in greenhouses. *(1 mark)*
2. Describe how you would control one of these factors in a commercial greenhouse throughout the year. *(2 marks)*
3. Explain how this factor is important in the growth of the plant and yield of tomatoes. *(2 marks)*

G–E

1 Light, or when you burn a paraffin heater which helps plants grow.
2 Add more light.
3 It is important to put lights on in the greenhouse for the plants.

Examiner: Whilst light is the name of a factor the second part of the answer is not a name, but the description of a process that might be carried out. No third factor has been named. All three factors are needed for the mark.

In this answer the candidate has not obeyed the command word. They have not described *how* to control the condition. There is no thought of the times of the year. No marks awarded.

This candidate has not given an explanation. They might have picked up on the wrong word as a command. The command word here is *explain*. Command words always start the question. This answer is more of a description. No marks awarded.

D–C

1 Light, warmth, and moisture.
2 I would control the temperature by using a heater in the greenhouse for the winter, when it is cold. Burning the fossil fuel will add heat. This is good because it will also add carbon dioxide to help the plants to grow.
3 If it is warm, but not too hot, the plants will grow at its best for us, because of more photosynthesis.

Examiner: Clear, but warmth is not as precise as temperature, as temperature is the specific name of a factor that can be controlled. Three factors identified.

Here the candidate has described the use of a method to control the temperature during the winter, and so has picked up on the command word. Unfortunately they have not covered the entire year, and not discussed temperature control in the summer. There is also considerable irrelevant information in the answer, which deviates from the factor, and from a simple description. 1 mark.

A fair answer. The candidate has explained that the factor is used in photosynthesis, although some of their grammar is not the clearest. This would have gained one mark. The second mark has been lost because the student has not linked the photosynthesis to increased yield.

B–A*

1 Light, temperature, and moisture.
2 You could control the amount of light in the greenhouse by using nets under the glass in the summer when the sun is bright, and by using electric lights during the dark winter days.
3 Light is needed for photosynthesis in the plant. The more light the plant gets the faster the rate of photosynthesis. Then there will be more tomatoes.

Examiner: Clear and precise. Three factors identified.

This is a good answer. The candidate has clearly read and dissected the question. The command word asks for a description of *how* to control the factor. The student has talked of using nets and lights. The student has also picked up on the fact that the question asks for the whole year, and talked about the action taken in summer and winter. They gain 2 marks.

The candidate has given an explanation that links the factor to photosynthesis, and increased yield. 2 marks.

Part 1 Course catch-up
Cells and the growing plant

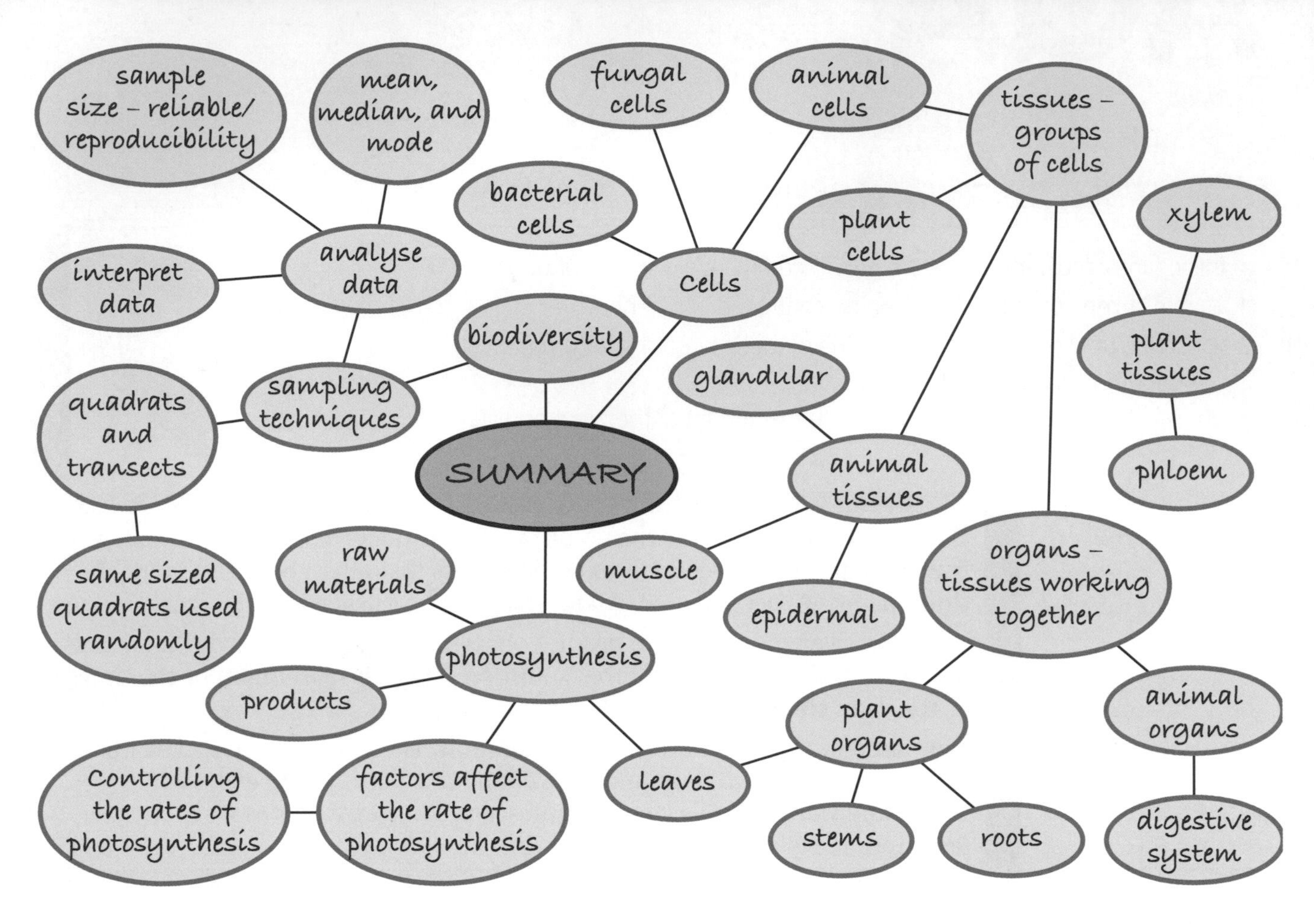

Revision checklist

- All living things are built out of cells.
- Cells contain parts that carry out different functions. Know the function of the nucleus, cytoplasm, cell membrane, cell wall, chloroplast, vacuole, ribosomes, and mitochondria.
- Know the difference between plant cells, animal cells, bacterial cells, and fungal cells.
- Cells become specialised for specific functions, for example, nerve cells can carry impulses.
- Diffusion is the process whereby molecules move into or out of cells. Diffusion is the random movement of molecules, and is affected by factors like distance, temperature, concentration gradient, and surface area.
- Similar cells work together in a tissue.
- Animal tissues include glandular, muscle, and covering. Plant tissues include xylem and phloem.
- Different tissues work together in an organ.
- Human organs include the stomach in which glandular, muscle, and covering tissue all work together. Plant organs include the leaves, stems, and roots.
- Different organs work together in an organ system to carry out a life process. The different organs might carry out different functions.
- An example of a human organ system is the digestive system. This system digests and absorbs food.
- Photosynthesis is the process by which plants make their own food. They use carbon dioxide from the air, light energy, and water from the soil, to make the waste product oxygen and the food glucose. Glucose can then be converted into other substances.
- The leaf is the organ in the plant involved with photosynthesis. It is highly adapted to carry out the function by being thin, having stomata for gaseous exchange, and having palisade cells with many chloroplasts.
- The rate of photosynthesis can be affected by light, temperature, and carbon dioxide levels. All of these factors can be controlled in a greenhouse.
- The distribution and biodiversity of organisms in the environment can be recorded by using sampling techniques. The numbers and location of the organisms can be measured using transect lines and regularly placed quadrats.
- Sampling techniques need to be valid by having a large sample size to make them repeatable and random or regularly placed sample locations to avoid bias. Often the data collected can be summarised by calculating a central value, such as a mean, mode, or median.

What is a protein?

Proteins are one of the major molecule groups that make up living things.

- They are built of **amino acids**.
- The amino acids are linked together in long chains.
- The chains are folded to give a specific shape.
- The shape is important for their function.
- The shape allows other molecules to fit into them.

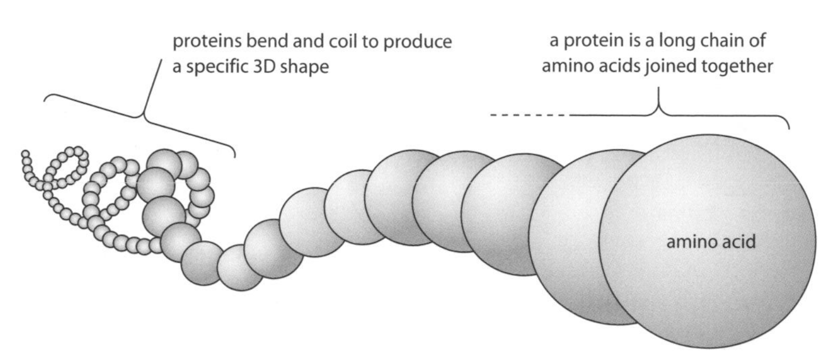

▲ Proteins are built from long chains of amino acids, which bend and coil into a specific shape.

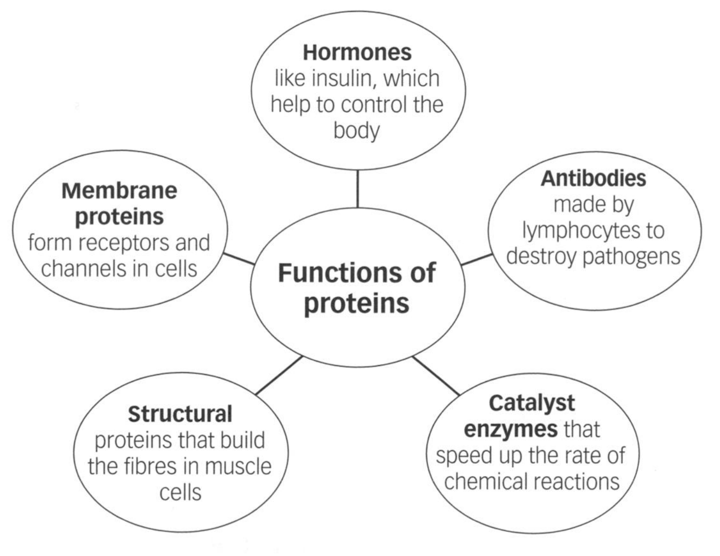

What are enzymes?

Enzymes are biological catalysts that speed up the rate of chemical reactions in the body.

- They are made of proteins.
- Without them the reactions of the body would be too slow for us to survive.
- The molecule that the enzyme works on is called the **substrate**.
- They can:
  - > break down large molecules into small ones, for example, in digestion
  - > build large molecules from small ones, for example, in photosynthesis.

Revision objectives

- ✔ know that proteins are made of long chains of amino acids
- ✔ understand a protein's shape allows it to carry out its function
- ✔ know that enzymes are proteins that catalyse chemical reactions
- ✔ understand that enzymes are specific, and work best at particular temperatures and pH

Student book references

2.14 Proteins

2.15 Enzymes

Specification key

✔ B2.5.2 a – b

Key words

protein, amino acid, hormone, antibody, catalyst, enzyme, substrate, specific, optimum, denatured

How enzymes work

The shape of the enzyme is vital for its function. The shape has an area into which substrate molecules can fit. This area is called the active site.

Substrate molecules fit into an enzyme's active site, for a reaction to occur. ▼

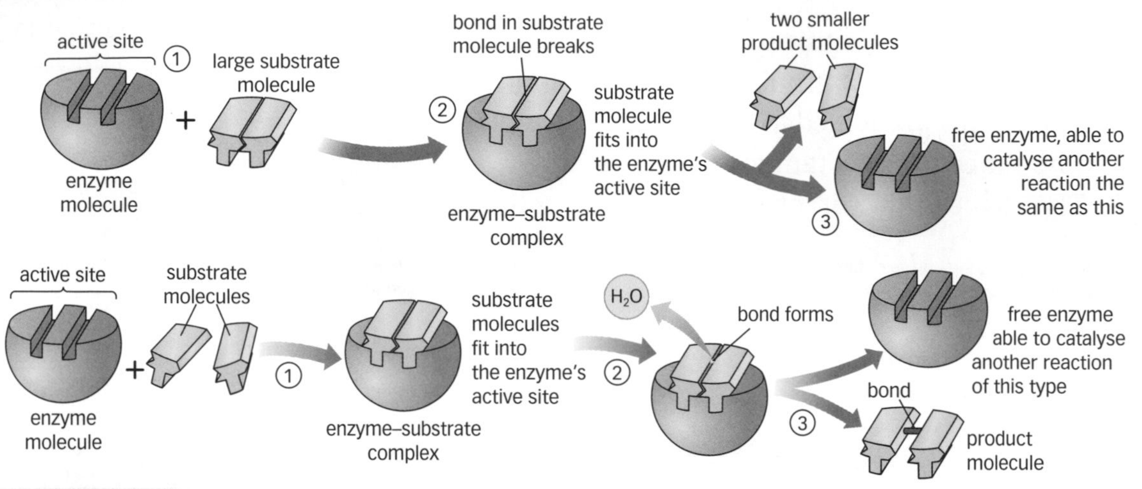

The key to the function of the enzyme is that the active-site shape is complementary to the substrate shape. This is not the same shape, but the two will fit together, like a key fits into a lock. No other substrate molecule will fit, which makes them **specific**.

Exam tip

AQA

Remember that a protein's shape is important for its function. This is often because the protein is complementary to another molecule, and the two will fit together. They are not the same shape.

What makes enzymes work best?

The way enzymes work is affected by temperature and pH.

Temperature

- As the temperature increases, the rate of reaction increases. This is because the temperature causes the enzyme and substrate to move more and bump into each other more often.
- This will not continue forever.
- Eventually the rate reaches a peak called the **optimum** temperature.
- Above the optimum, the increase in temperature starts to damage the shape of the enzyme. It cannot then work. The enzyme is said to be **denatured**.

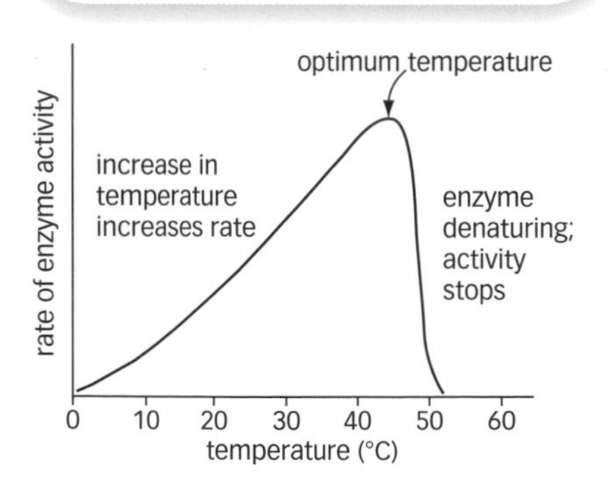

▲ Graph to show the effect of temperature on the rate of enzyme reactions.

pH

- Each enzyme has an optimum pH. Here it works best.
- Above or below this level it does not work so well. This is because the shape of the enzyme active site is damaged.
- It is denatured.

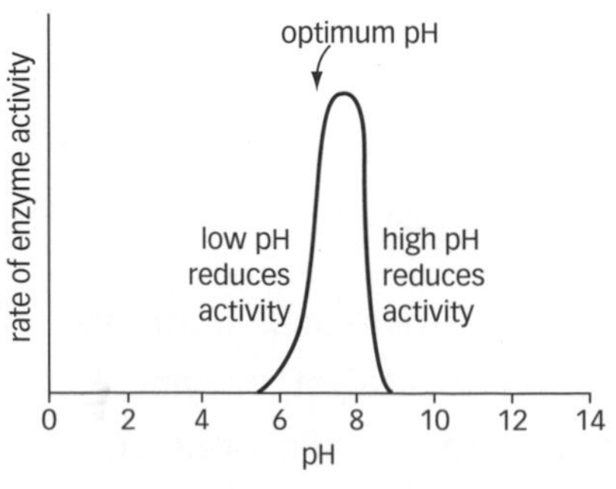

▲ Graph to show the effect of pH on the rate of enzyme reactions.

Questions

1 How are proteins constructed?

2 What **five** types of proteins function in the body?

3 **H** How do enzymes work?

Digestion

Digestion is the breakdown of large insoluble food molecules into small soluble molecules that can be absorbed into the blood. This provides the nutrients for us to survive.

Enzymes are important in digestion.

- They catalyse these breakdown reactions in the gut.
- They are produced by specialised cells in glandular tissue.
- They are released into the cavity of the gut.
- They then break down the food inside the gut.

Revision objectives

- know that the enzymes used in digestion are produced in glands, and work outside the cell
- understand where the main digestive enzymes are produced, and how they act
- know the role of hydrochloric acid and bile in helping digestion

Student book references

2.16 Enzymes and digestion

Specification key

✔ B2.5.2 c – h

The digestive process

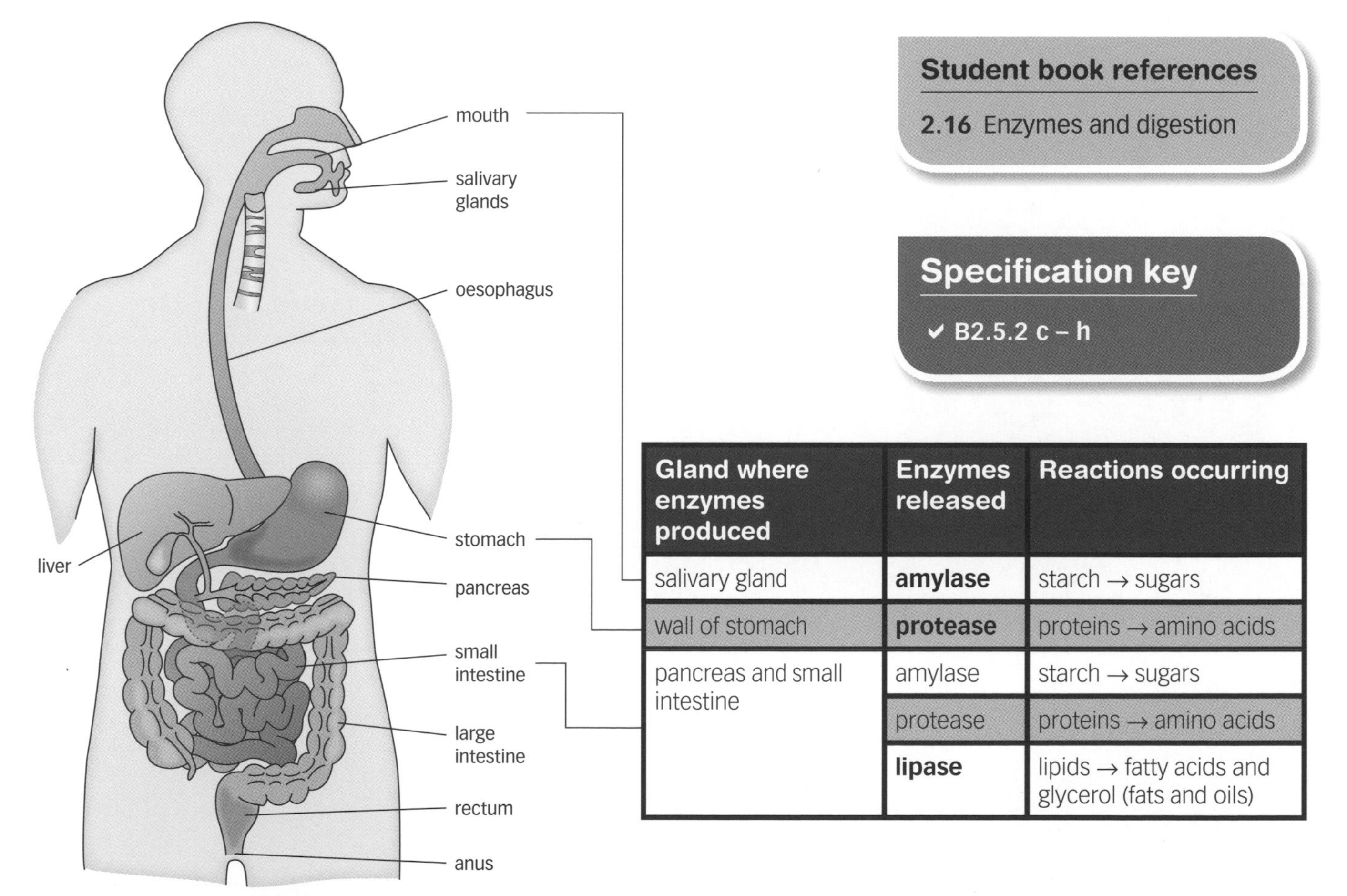

Gland where enzymes produced	Enzymes released	Reactions occurring
salivary gland	**amylase**	starch → sugars
wall of stomach	**protease**	proteins → amino acids
pancreas and small intestine	amylase	starch → sugars
	protease	proteins → amino acids
	lipase	lipids → fatty acids and glycerol (fats and oils)

▲ Enzymes in the digestive process.

Key words

digestion, amylase, protease, lipase, hydrochloric acid, bile

Exam tip

Learn the three main digestive enzymes, where they are made, where they act, and what they do.

The importance of pH

The pH varies in the gut, and this affects which enzymes can work.

Acid pH in the stomach

The wall of the stomach produces **hydrochloric acid**. Only the stomach protease can work in this pH. The acid also helps by killing bacteria that enter the gut.

Neutralising the acid in the small intestine

The acidic food entering the small intestine from the stomach is neutralised by **bile**.

Bile is made in the liver → stored in the gall bladder → released into the small intestine → neutralises the acid.

This results in slightly alkaline conditions, which are best for the enzymes in the small intestine.

Questions

1 What is the process where food is broken down in the gut called?

2 What does amylase do?

3 What is the relationship between pH and enzyme action in the gut?

Working to Grade E

1 What are proteins made of?

2 Draw a diagram to show how the molecules that make proteins are arranged together.

3 Define an enzyme.

4 What is an enzyme made of?

5 Define digestion.

6 Bile is a digestive juice.

  a Where is bile made?

  b Where is bile stored?

Working to Grade C

7 Why is the structure of a protein important?

8 Complete the following table.

Type of protein	Function of protein
	Bonds to pathogens, destroying them.
Enzyme	
	Allows substances into cells through membranes.
	Controls the body's functions.

9 Give an example of structural proteins in the body.

10 Place the following statements into the correct order to explain the action of an enzyme.

  a The substrate molecules are brought together to form a bond.

  b The product leaves the active site.

  c Substrate joins the enzyme at its active site.

  d Once the bond is formed this is called the product.

  e The enzyme is once again available to perform another reaction.

11 What is the role of enzymes in:

  a digestion?

  b photosynthesis?

12 Temperature affects the rate of an enzyme-controlled reaction. What other factor will affect the rate of enzyme controlled reactions?

13 Why are enzymes vital for the survival of an organism?

14 Look at the table below and complete the blanks.

Region of enzyme action in the gut	Enzymes released	Reactions occurring
	amylase	starch → sugars
stomach		proteins → amino acids
		starch → sugars
	protease	
		lipids → fatty acids and glycerol (fats and oils)

15 How does hydrochloric acid help in the process of digestion?

16 How does the pancreas aid digestion?

Working to Grade A*

17 Explain what makes enzymes very specific.

18 Look at the graph below, which shows the change in the rate of an enzyme controlled reaction as the temperature changes.

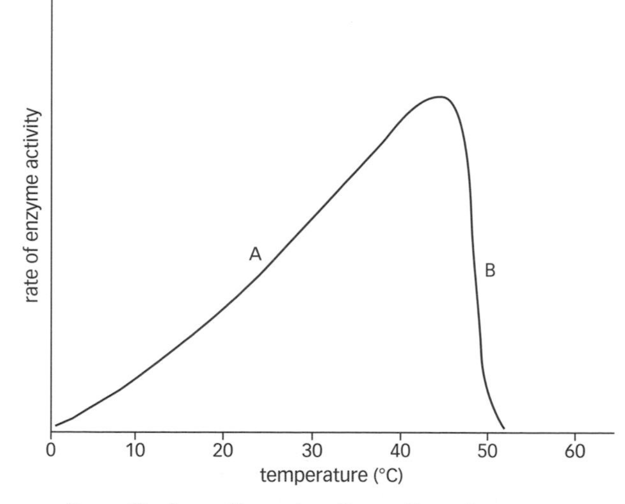

  a Describe how the rate of reaction changes as the temperature increases.

  b Explain why the rate is changing at point A.

  c Explain why the rate is changing at point B.

19 What is the role of bile in digestion?

Examination questions
Enzymes and digestion

1 Enzymes are involved in digestion. There are two types of protease involved in the digestion of proteins, but they work in different areas of the digestive system.

a What is the function of an enzyme?

....

....

(1 mark)

b Look at the graph, which shows the rate of the two different enzymes at different pH values.

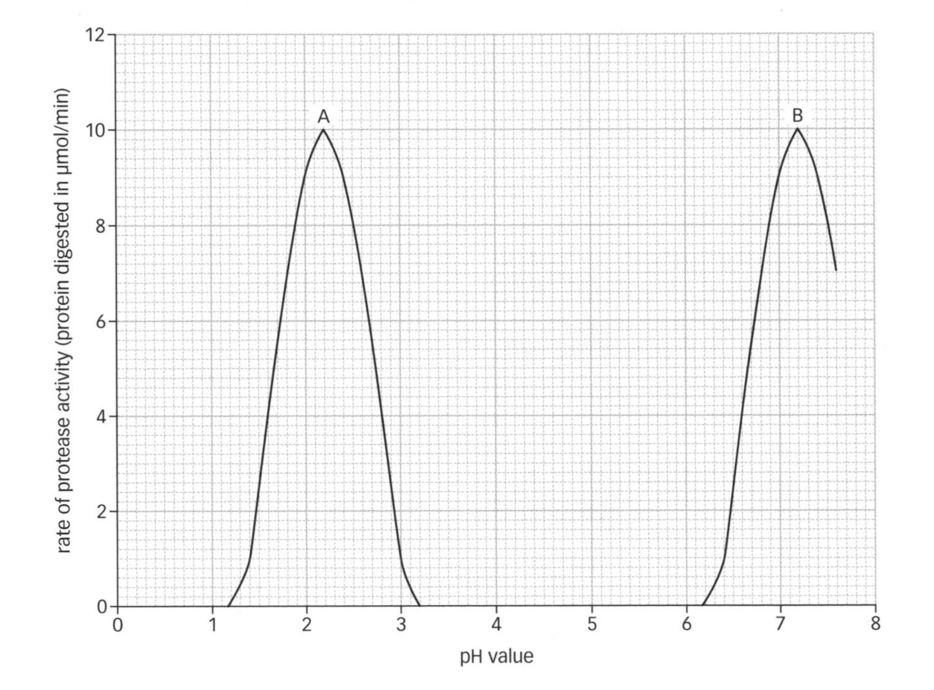

i In which region of the digestive system will enzyme A work?

....

(1 mark)

ii Explain your reasons for this answer.

....

....

....

(2 marks)

iii Complete the line for enzyme B to show how the rate changes as the pH increases.

(1 mark)

iv The optimum pH is the rate where the enzyme reaction is greatest. What is the optimum pH for enzyme A?

....

(1 mark)

c Enzyme A does not work at the higher pH values of enzyme B. Explain in detail what has happened to the enzyme to prevent it working at this higher pH.

....

....

....

....

....

....

....

....

....

....

(4 marks)

(Total marks: 10)

10: Applications of enzymes

Revision objectives

- know that enzymes from microorganisms are used in biological detergents
- understand the advantages and disadvantages of using enzymes in industry
- be aware of examples of the use of enzymes in industry

Student book references

2.17 Enzymes in the home – detergents

2.18 Enzymes in industry

Specification key

✔ B2.5.2 i – j

Obtaining commercial enzymes

Enzymes speed up the rate of chemical reactions. So they can be very useful in:

- the home – such as in washing powders
- industry – such as in food production.

We are able to obtain many of these enzymes from microorganisms grown in fermenters. The microbes make the enzymes and release them from their cells. We can then collect them for our use.

Biological washing powders

Detergents are cleaning agents. Detergents like washing powders are used to remove stains from clothes. Biological washing powders contain enzymes to help in the removal of stains. The advantage of these powders is that they remove stains that other powders leave behind.

stains on clothes are made mainly of proteins and fat

lipases digest fat stains into fatty acids and glycerol, which also wash away

proteases in the powder digest proteins into amino acids that wash away

▲ How enzymes work to remove stains.

No need to boil wash

Biological washing powders have a second great advantage. They can remove stains at lower temperatures than other washing powders. Most enzymes work best at 40°C. So these powders will remove stains at these low temperatures and there is no need to boil wash. We can now wash more delicate fabrics, and save energy as well.

Enzymes in industry

Industrial processes are expensive because of the conditions they require. Enzymes are useful in industry because they allow reactions to occur:

- at lower temperatures
- at lower pressures.

This reduces the need for expensive, energy-demanding equipment.

Uses of enzymes

There are many examples of the use of enzymes. Different industries use different enzymes.

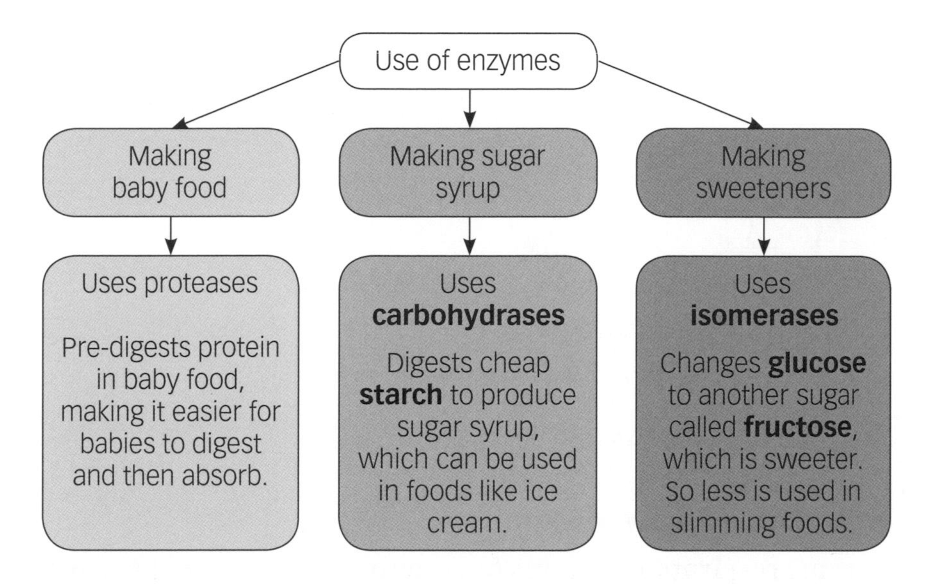

Disadvantages of using enzymes

Unfortunately it's not all good news! There are a few disadvantages with using enzymes in industry.

- The enzymes will be denatured at high temperatures. This means that their shape has changed and they can no longer work.
- They can be expensive to produce. This means that they need to save the industry a lot of money to be worth using at all.

Key words

enzyme, detergent, protease, carbohydrase, starch, isomerase, glucose, fructose

Exam tip AQA

For each use of the enzymes there are advantages and disadvantages. These are all linked to the characteristics of enzymes learnt in B2 9.

Questions

1 What is the purpose of using enzymes in washing powders?

2 Name three enzymes used in industry.

3 **H** Explain why the temperature of industrial processes using enzymes needs to be maintained at 40 °C.

Revision objectives

- know that respiration is the release of energy, and occurs aerobically and anaerobically
- understand the process of aerobic respiration
- understand the uses of energy in plants and animals
- know the changes that occur in our bodies to allow energy release during exercise
- understand the process of anaerobic respiration

Student book references

2.19 Energy and life processes
2.20 Aerobic respiration
2.21 Anaerobic respiration

Specification key

✔ B2.6.1 ✔ B2.6.2

Key words

respiration, aerobic, anaerobic, mitochondria, exercise, glycogen, lactic acid, fatigue, oxygen debt

Respiration

Respiration is the process where cells release energy from molecules like sugar. It occurs continuously in both plants and animals.

It is controlled by enzymes. It can occur in two ways:

- **aerobic** – with oxygen
- **anaerobic** – without oxygen.

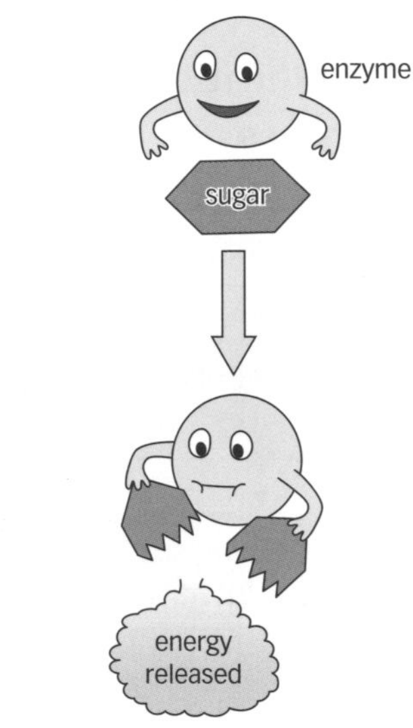

▲ Energy release from a sugar molecule.

Aerobic respiration

Aerobic respiration requires oxygen to release the energy from sugars like glucose.

- It is very efficient, releasing a lot of energy.
- The reactions of aerobic respiration occur mainly in tiny structures in the cell called **mitochondria**.

The equation for aerobic respiration is:

glucose + oxygen → carbon dioxide + water (+ energy)

How is the energy used by the organism?

Use	Explanation
Building larger molecules	Energy is used to build small molecules into larger ones. For example, amino acids are built into proteins in both plants and animals. Plants add nitrates to sugars and other nutrients to build amino acids.
Muscle contraction in animals	Energy is needed to cause the muscle proteins to contract. This will bring about movement.
Maintaining body temperature in mammals and birds	Energy is released as heat, which helps to keep the bodies of birds and mammals at a constant temperature. This keeps them active in colder surroundings.

Energy and exercise

Large amounts of energy are required in our bodies when we **exercise**. To release more energy a number of changes occur in our body.

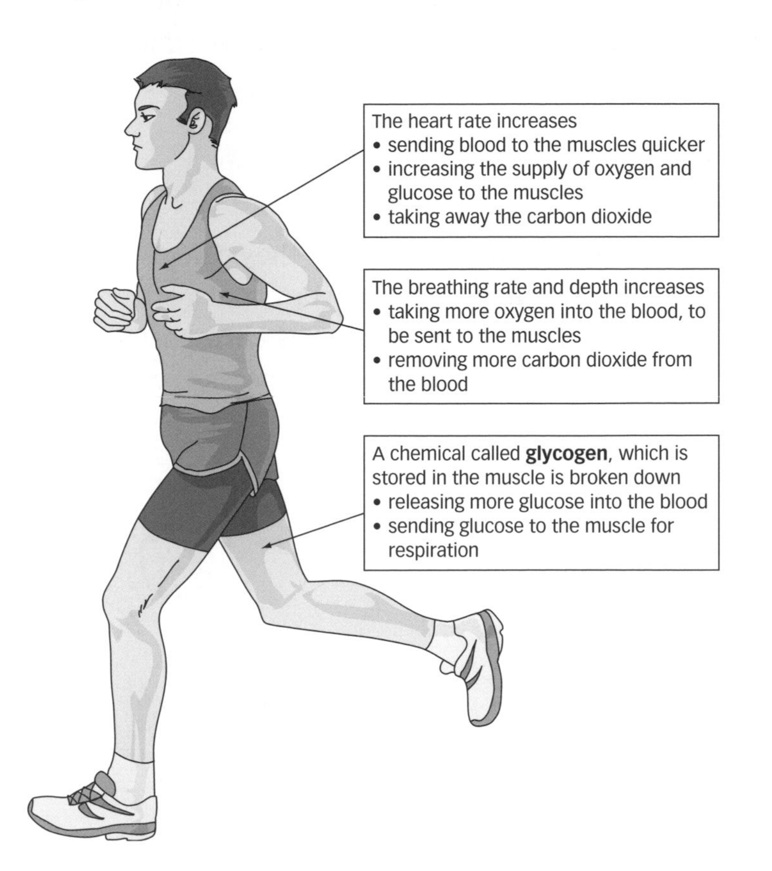

Anaerobic respiration

This type of respiration only occurs when there is not enough oxygen.

- It is less efficient at releasing energy than aerobic respiration.
- This is because it is an incomplete breakdown of glucose.
- The reactions occur in the cytoplasm of cells.
- This happens in muscles during intense or sprinting activities.
- The waste product is **lactic acid**.

Glucose → lactic acid (+ a little energy)

Lactic acid is toxic and it builds up in muscles during long periods of vigorous exercise. When it reaches high levels it causes the muscles to become **fatigued**. This means they will no longer contract efficiently.

The blood flowing through the muscles will remove the lactic acid and take it to the liver where it will be broken down.

Exam tip 

Learn the equation for aerobic respiration. It is a common exam question. Most questions on exercise and respiration relate to the four components of the equation.

H Oxygen debt

The only way to get rid of the toxic lactic acid is to use oxygen. So any build-up of lactic acid means the body needs oxygen; this is called **oxygen debt**.

The lactic acid produced in anaerobic respiration lowers the blood pH. This will cause an increase in breathing rate. This results in more oxygen being taken into the body, repaying the oxygen debt. The oxygen goes to the liver and the lactic acid is broken down into carbon dioxide and water.

Questions

1. Where does respiration occur?
2. What is the difference between aerobic and anaerobic respiration?
3. During exercise a number of changes happen in the body to help cells release more energy. Explain how **two** of these changes help in energy release.

12: Cell division

Revision objectives

- know that mitosis is a type of cell division that produces identical body cells
- know the uses of mitosis
- know that meiosis produces gametes, which have half the chromosome number
- know the role of meiosis, and where it occurs

Student book references

2.22 Cell division – mitosis

2.23 Cell division – meiosis

Specification key

✓ **B2.7.1 a – i, n**

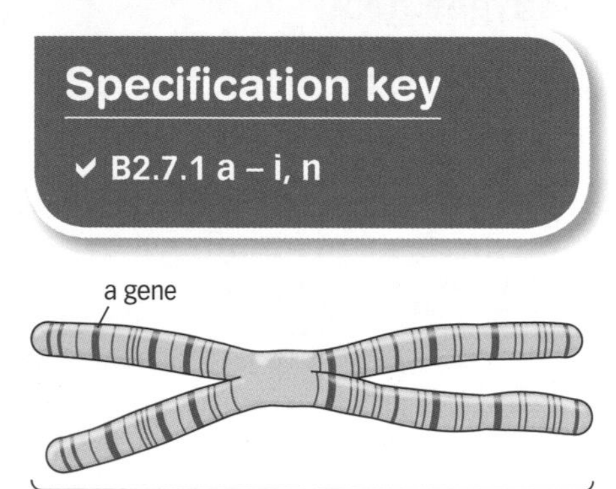

▲ Genes on a chromosome.

Exam tip

AQA

Questions often expect you to know the number of chromosomes in the new cells produced by either mitosis or meiosis. Remember in mitosis the number stays the same, in meiosis the number halves.

What are chromosomes?

Chromosomes are thread-like structures in the nucleus of every cell. Each chromosome contains many pieces of information called **genes**. Chromosomes are made of a chemical called DNA. In body cells chromosomes are found in pairs.

Mitosis

Body cells divide by **mitosis**. Before they can divide, each chromosome must make an exact copy of itself so that there will be one copy for each new cell. The chromosome then has an 'X' shape (as shown below). This process is called DNA replication.

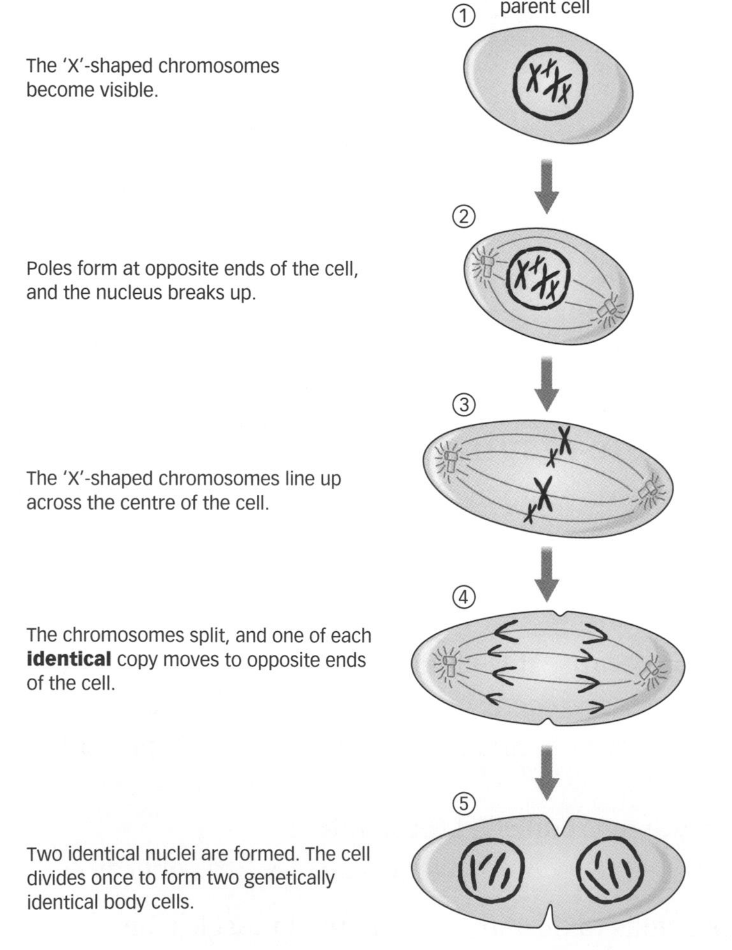

Where does mitosis occur?

Mitosis occurs for:

- growth – new cells cause the body to get bigger
- repair – to replace old or damaged cells
- asexual reproduction – to produce new individuals, genetically identical (contain the same alleles) to each other and to their parent, called clones.

Gametes

Body cells always have chromosomes in pairs. During sexual reproduction **gametes** are made.

- Eggs are made in the ovaries of females.
- Sperm are made in the testes of males.

Gametes will only have one of the pair of chromosomes. To produce cells with only half the number of chromosomes, a second type of division is used, called **meiosis**.

Key words

chromosome, gene, mitosis, identical, gamete, meiosis, fertilise

H Meiosis

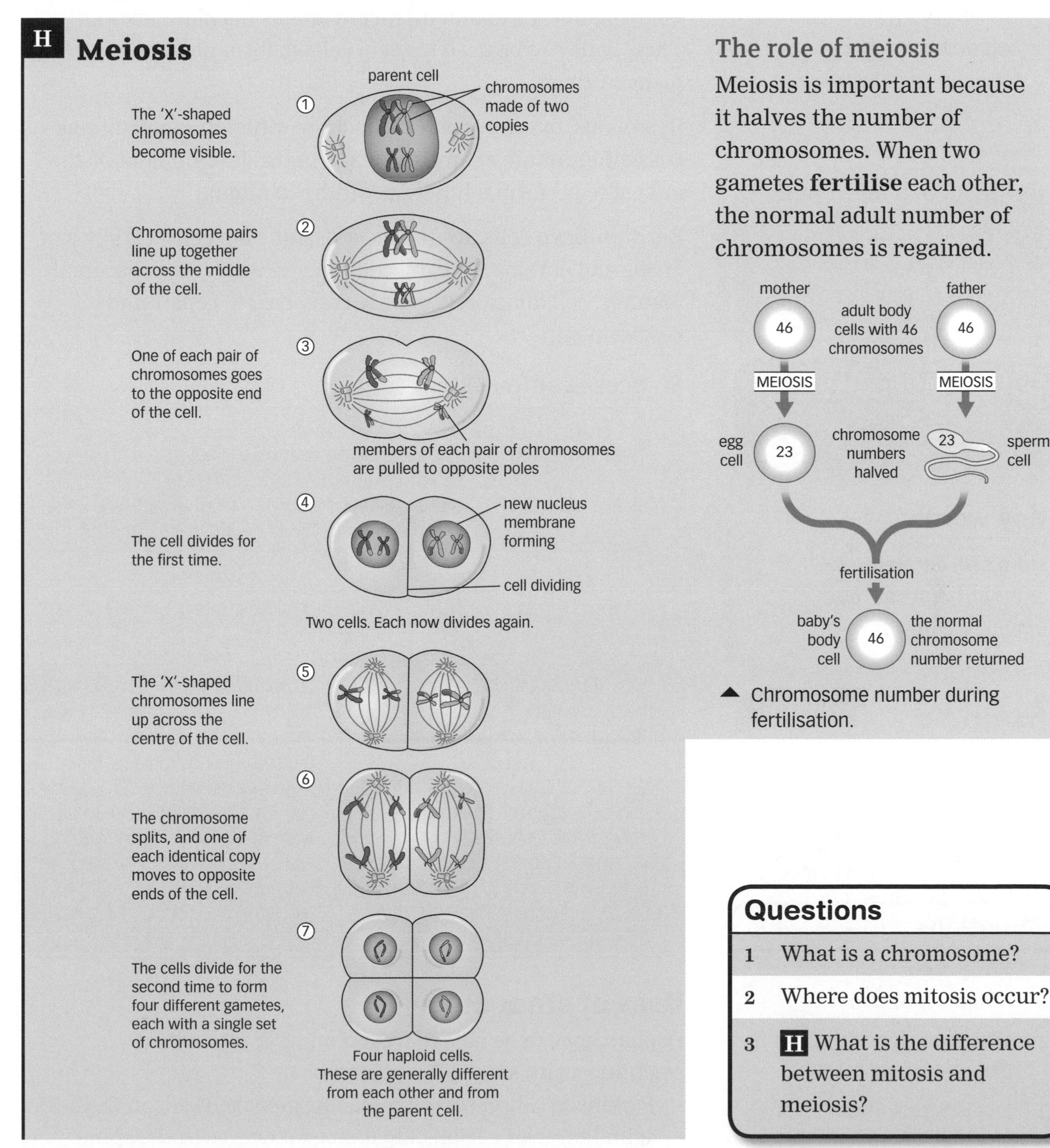

The role of meiosis

Meiosis is important because it halves the number of chromosomes. When two gametes **fertilise** each other, the normal adult number of chromosomes is regained.

▲ Chromosome number during fertilisation.

This newly formed cell created through meiosis then repeatedly divides by mitosis to develop into a new individual.

Questions

1 What is a chromosome?

2 Where does mitosis occur?

3 **H** What is the difference between mitosis and meiosis?

13: Stem cells

Revision objectives

- know what a stem cell is
- know the sources of stem cells in plants and animals
- know that stem cells can be used to treat medical conditions
- appreciate the ethical concerns with the use of stem cells

Student book references

2.29 Embryo screening, stem cells, and DNA fingerprinting

Specification key

- B2.7.1 j – m

Key words

stem cell, differentiate, embryo, bone marrow, umbilical cord

Exam tip

This topic is quite straightforward. You need to know **what** stem cells are, **where** they are found, and **how** they can be used.

Questions

1. What does undifferentiated mean?
2. Which source of stem cells is the most useful?
3. Explain **one** ethical issue associated with the use of stem cells.

What are stem cells?

Stem cells are undifferentiated cells. That means that they have not specialised as any one type of cell. These cells are useful to scientists because they can make them divide and differentiate into cell types that they need.

Where do we get stem cells from?

Many plant cells do not **differentiate** as the plant grows. These cells can be used as stem cells to form new roots on plant cuttings.

In animals, finding stem cells is more difficult. Most animal cells differentiate at an early stage in the development of the animal to perform a function within an organ.

Early **embryo** cells are the most useful stem cells as they can divide and develop into any cell type we want. As the animal matures, cell division is mainly restricted to repair and replacement.

Sources of animal stem cells

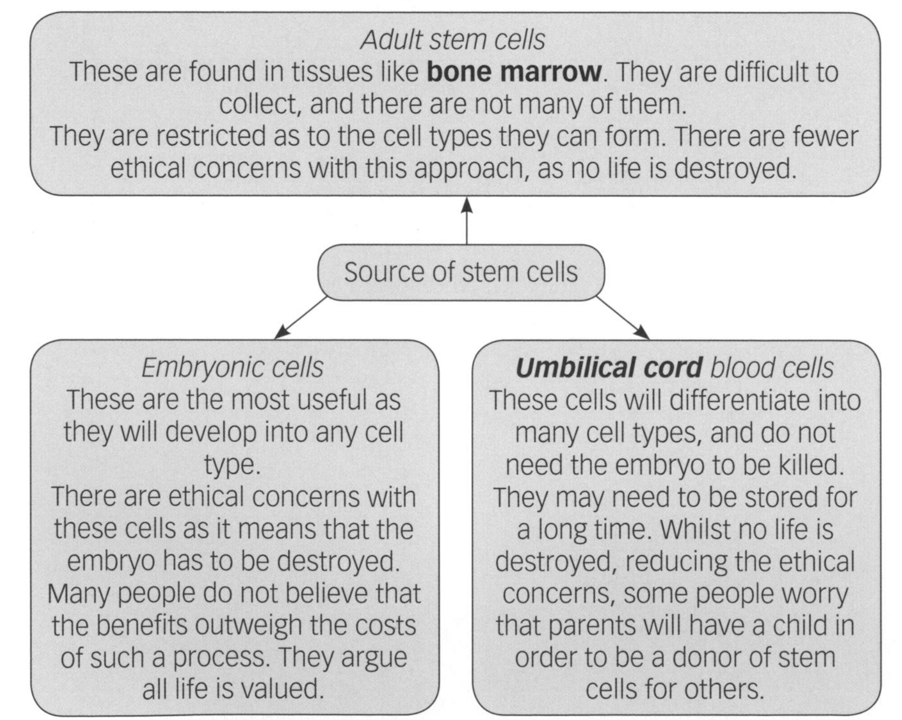

Uses of stem cells

Doctors hope to be able to treat a number of medical conditions with stem cells, such as:

- Parkinson's disease – by replacing damaged cells in the brain
- spinal injuries – replacing damaged nerves in the spine
- organ creation – for transplant
- diabetes – replacing damaged pancreas cells.

Questio
Enzyme

Working to Grade E

1 What is a detergent?
2 What **two** enzymes are used in washing powders?
3 The following statements refer to the advantages of using enzymes in industrial processes. For each statement, complete the sentence by using the word higher or lower as appropriate.
  a The reactions can be carried out at _______ temperatures.
  b The reactions can be carried out at _______ pressures.
  c The reactions will occur at a _______ rate.
  d The cost of the process will be _______ .
4 Define cellular respiration.
5 What is the source of energy in respiration?
6 Give **three** uses of the energy released by respiration in the body.
7 Energy is used during exercise. A number of changes occur in the body of an athlete to ensure that they can release energy. What happens to the athlete's heart rate during exercise?
8 There are two types of cell division: mitosis and meiosis. Which type of division is used:
  a when the body is growing?
  b to make gametes?
  c to repair damaged tissues?
9 Where are the gametes produced in males and females?

Working to Grade C

10 What substrates do the enzymes in washing powders work on?
11 What are the advantages of putting enzymes into washing powders?
12 Why are biological washing powders not effective in a boil wash?
13 Enzymes are used in industry.
  a For **each** of the following food products identify the:
    i substrate used
    ii enzyme
    iii product made.

  b Name a disadvantage of using enzymes in industry.

14 There are two types of respiration: aerobic and anaerobic. For **each** process, make a list of:
  a the reactants
  b the products.
15 There are two types of cell division: mitosis and meiosis. Which type of division is used during asexual reproduction?
16 Look at this drawing of an animal cell with four chromosomes.

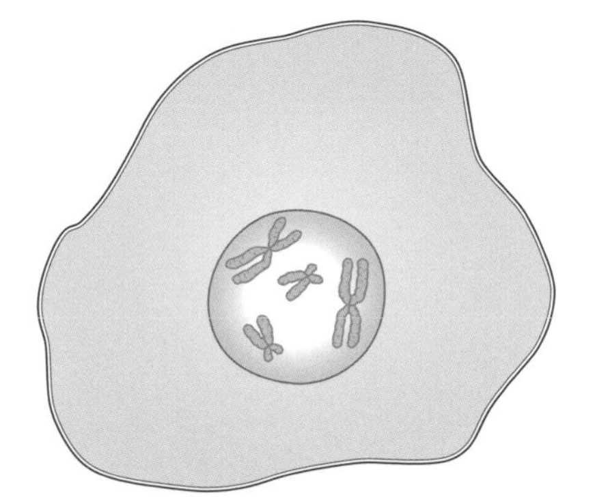

  a How many chromosomes will the daughter cells have if the cell divides by mitosis?
  b How many chromosomes will the daughter cells have if the cell divides by meiosis?
  c How many daughter cells will be produced if the cell divides by meiosis?
17 Name **two** sources of animal stem cells.
18 What is a stem cell?
19 Name a condition that stem cells may be used to treat in the future.
20 What is differentiation?

Working to Grade A*

21 Explain the reason why an enzyme is used in making the following products:
  a baby food.
  b slimming bar.
22 Explain how carbohydrases are used in the food industry.
23 Look at question 7, above. Explain why the athlete's heart rate changes during exercise.
24 What is oxygen debt?
25 What does the body do to relieve oxygen debt?

Examination questions
Enzymes, respiration, and cell division

1 Respiration occurs in cells, when they release energy. Complete the following diagram to show what the cell needs, and the waste products it produces during respiration.

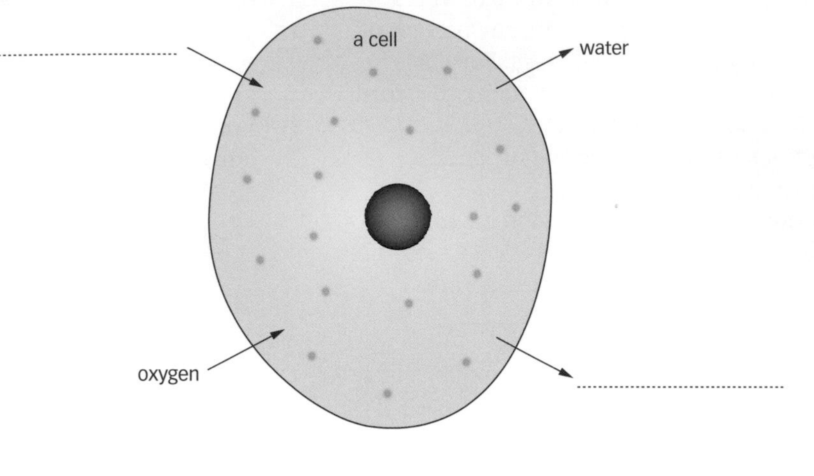

(2 marks)
(Total marks: 2)

2 Put the following parts in rank order of size, starting at the smallest.

Cell **Gene** **Nucleus** **Chromosome**

…………………… → …………………… → …………………… → ……………………

(3 marks)
(Total marks: 3)

3 Below is a diagram of a type of cell division, showing the number of chromosomes in some of the cells.

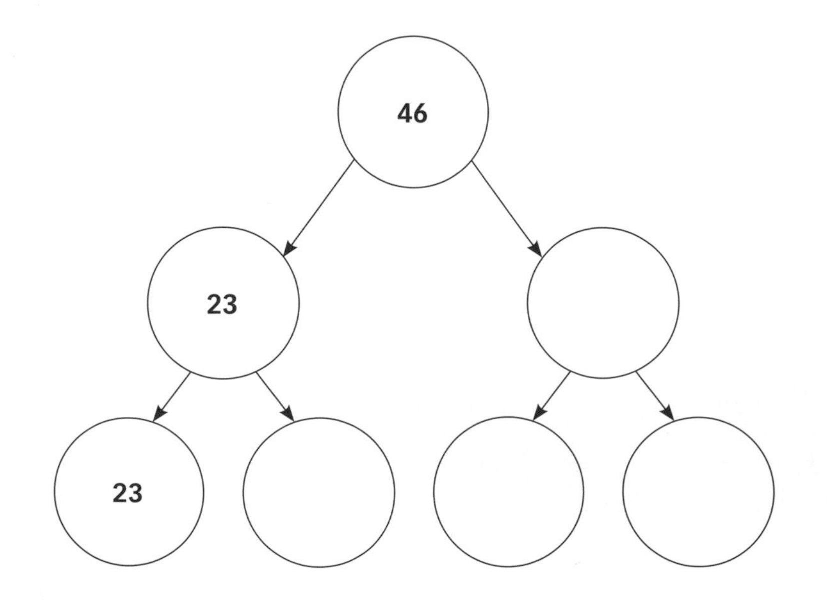

a What type of division is this called?

……………………………………………………………………………………………………

(1 mark)

b Complete the diagram to show how many chromosomes are found in the other cells. *(1 mark)*

c What type of cell is produced by this type of division?

....

....

(1 mark)

(Total marks: 3)

4 Stem cells are cells that can be grown in the laboratory, and are used by doctors to treat a number of conditions.

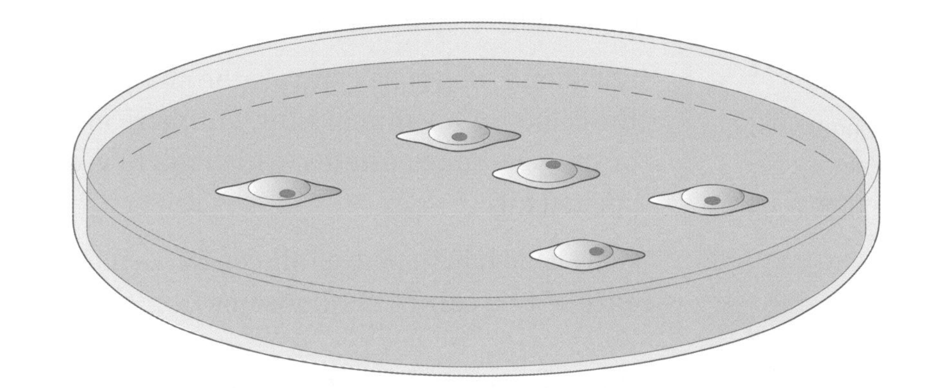

a Why are stem cells so useful to doctors?

....

....

....

(1 mark)

b Name **one** source of stem cells.

....

(1 mark)

c Some people have expressed concern about the use of stem cells. Outline **one** argument that people can put forward against the use of stem cells.

....

....

....

(1 mark)

(Total marks: 3)

Revision objectives

- understand that there are different forms of genes called alleles
- be able to follow monohybrid inheritance
- be familiar with the work of Mendel on inheritance
- understand that the rediscovery of Mendel's work helped us to explain the processes of inheritance

Student book references

2.24 Inheritance – what Mendel did

2.25 The importance of Mendel's work

Specification key

- B2.7.2 a, c – e

Genes

Each individual is unique. They are a mix of characteristics from the father and mother. Characteristics are controlled by **genes**. Each gene is a section of DNA on a chromosome in the nucleus. Human body cells have 23 pairs of chromosomes. Eggs and sperm only have one of each pair.

- We inherit half our genes from our mother in the egg.
- We inherit half our genes from our father in the sperm.

Sometimes one gene controls the characteristic, for example, the colour of an animal's fur, but there are different versions of the gene called **alleles**, which might give the animal spotted fur.

Body cells have a pair of alleles for a characteristic, one on each of one pair of chromosomes.

Tracking inheritance

Inheritance is the passing of characteristics from one generation to the next. This means that genes are passed on. It is possible to follow the gene from one generation to the next. Following one characteristic like this is called monohybrid inheritance. Take this example of an animal's fur colouring:

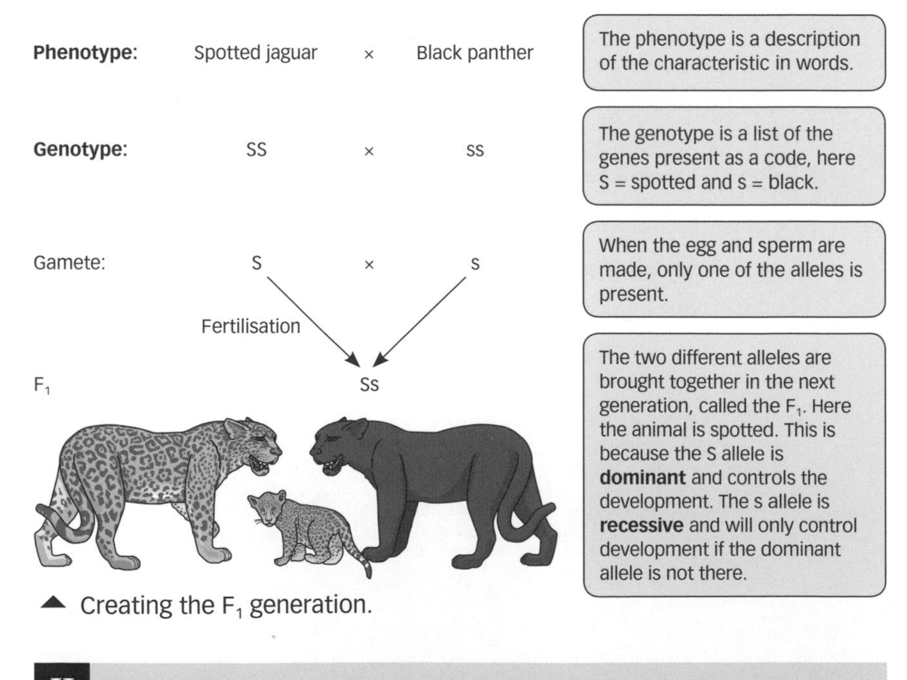

Creating the F_1 generation.

H Two important words to know here are:

- **Homozygous** – here the genotype has identical alleles, for example, SS.
- **Heterozygous** – here the genotype has different alleles, for example, Ss.

Mating (crossing) the F_1 leopards

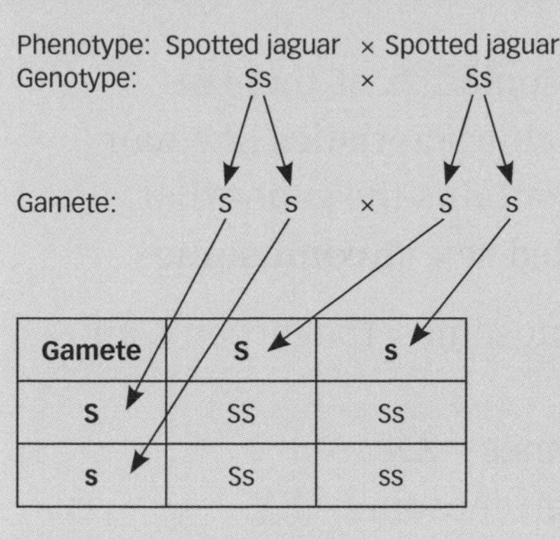

Gamete	S	s
S	SS	Ss
s	Ss	ss

The genotype is heterozygous.

Each parent produces two different types of gamete. One has the S allele, the other has the s allele.

This is called a Punnet square. It allows us to create all possible combinations of the gametes. This allows us to predict the possible genotypes/phenotypes in the offspring in the second generation, called the F_2.

The F_2 results are:

3 spotted jaguars SS Ss Ss and 1 black panther ss

Try predicting the result of crossing a spotted jaguar (Ss) with a black panther (ss).

▲ Creating the F_2 generation.

Key words

gene, allele, phenotype, genotype, dominant, recessive, homozygous, heterozygous, Mendel, factor

Mendel – the father of inheritance

Gregor **Mendel** was born in 1822. He made a number of startling observations, which form the basis of our understanding of genetics. However, the importance of his work was not realised until after his death.

Mendel carried out many breeding experiments on pea plants, controlling the transfer of pollen from one plant to another. He was controlling the crossing of alleles, although he did not realise it. He worked at a time when scientists had not discovered chromosomes and certainly had not linked inheritance to them. As there was a lack of scientific knowledge at the time, the importance of Mendel's work was overlooked.

- He thought that inheritance of a characteristic was controlled by **factors** (we now call these genes).
- He worked out that the 'factors' must be in pairs in the adult cells.
- Only one of the factors would be in the gamete.
- The offspring would contain two factors, one from each parent.
- He was able to predict the outcome of crosses, just as we have shown above.
- The ratio of 3:1 in the F_2 is now called a Mendelian ratio.
- Finally his work was understood about 40 years later when other scientists could explain his results using genes.

Exam tip AQA

If you are working towards Foundation, you just need to be able to understand and interpret the genetic diagram. If you're working towards Higher, you should be able to construct them for yourself. Treat genetics questions like a maths question: practise with different genes.

Questions

1 What is inheritance?

2 Who was the pioneer of genetics?

3 **H** What genetic cross will always produce a 3:1 ratio in the offspring?

15: Genes and genetic disorders

Revision objectives

- understand how sex is inherited
- know the structure and function of DNA
- know that DNA fingerprinting is a method for identifying people
- know about some genetic disorders that are inherited

Student book references

2.26 The inheritance of sex
2.27 How genes control characteristics
2.28 Genetic disorders
2.29 Embryo screening, stem cells, and DNA fingerprinting

Specification key

- B2.7.2 a – b, f – i
- B2.7.3

The inheritance of sex

Humans have 23 pairs of chromosomes; 22 of them are matching pairs and control body characteristics like hair colour and eye colour. One pair contains the genes that determine our sex. These are called **sex chromosomes**.

There are two different sex chromosomes, the larger X and smaller Y chromosomes.

- Females have two X chromosomes – XX
- Males have one X and one Y chromosome – XY

Every time humans reproduce there is a 50:50 chance of having a boy or a girl, as shown in the diagram below.

Parents	Male	×	Female
Chromosomes	XY	×	XX
Gametes	X Y	×	X X

Gametes	X	X
X	XX	XX
Y	XY	XY

½ girls
½ boys

How does DNA work?

As we have learnt, chromosomes are made of a chemical called deoxyribonucleic acid (DNA). DNA is a spiral molecule, 'like a twisted ladder' called a double helix. Genes are just small sections of the DNA.

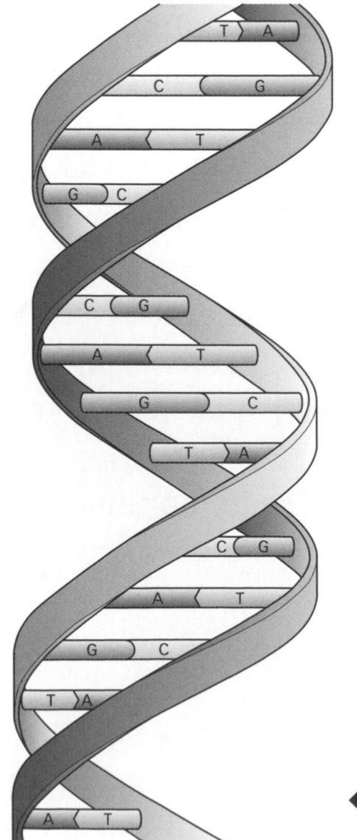

Simplified diagram of a DNA molecule.

H The genetic code

Each gene works by coding for the sequence of amino acids in a **protein**.

- The genetic code is contained in a sequence of **bases** in the DNA molecule.
- There are four different bases: A, T, C, and G.
- Each gene is made of hundreds of bases in a sequence.
- The cell reads the base sequence, three bases at a time.
- Each set of three bases is called a triplet.
- Each triplet codes for a specific amino acid.
- Each specific amino acid is joined to the next, and gradually builds a protein.
- The order of the amino acids is determined by the sequence of bases in the DNA.

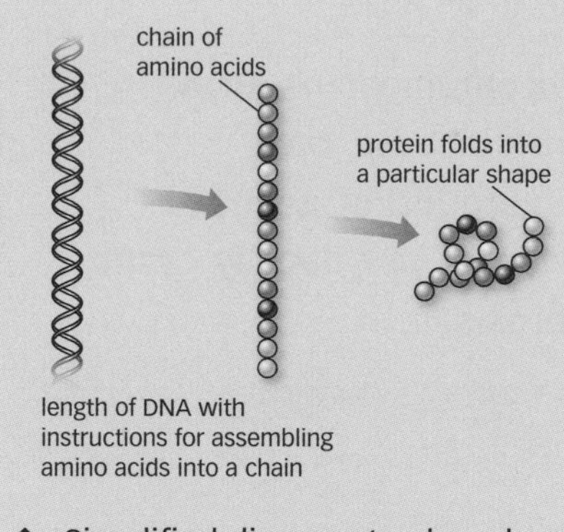

▲ Simplified diagram to show how the coded information in a gene determines the shape and the function of a protein.

DNA fingerprinting

Everyone's DNA is unique, apart from identical twins. This fact can be used to identify individuals. The technique of **DNA fingerprinting** was developed in the 1980s and has two major uses:

- to establish family connections like paternity
- to identify a criminal from evidence found at a crime scene.

The DNA sample collected is cut by enzymes, and separated to produce a series of bands on a gel. Scientists just need to match the bands to make an identification. Care is needed in the technique to avoid contaminating the sample.

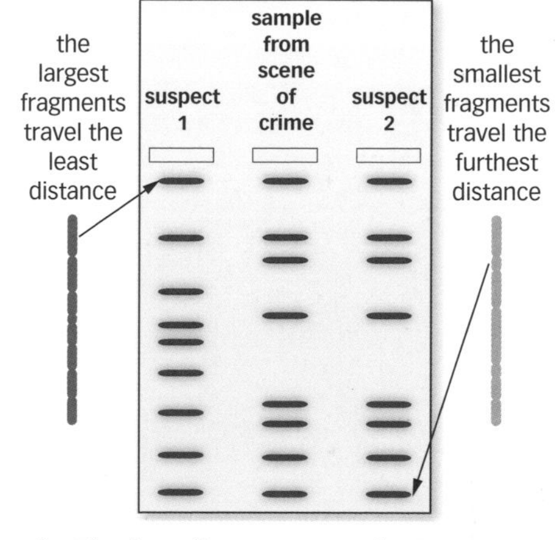

▲ The banding pattern of a DNA fingerprint.

Inherited disorders

Sometimes a gene has a defect that results in a **genetic disorder**. There are about 5000 genetic disorders. Two examples are **polydactyly** and **cystic fibrosis**.

Polydactyly

This is a condition where additional digits develop on the hands or feet. It is caused by a dominant allele, and can then be passed on by one parent who has the disorder.

Cystic fibrosis

This is a disorder of the cell membranes. It results in a mucus build-up in the lungs. It can make sufferers more vulnerable to chest infections. Other organs like the pancreas can be affected, which might affect the digestive process.

Key words

sex chromosome, DNA, protein, base, DNA fingerprinting, genetic disorder, polydactyly, cystic fibrosis

It is caused by a recessive allele (c); the normal allele would be dominant (C). So three possible genotypes can exist:

- normal individual – CC
- carrier (no symptoms, but with the allele) – Cc
- sufferer – cc.

To be a sufferer the allele must be inherited from both parents.

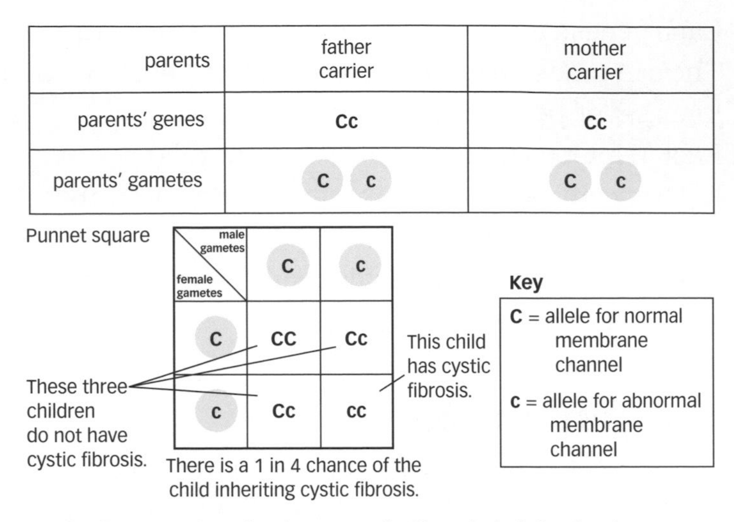

▲ A genetic diagram showing how cystic fibrosis is inherited.

Embryo screening

If both parents carry the cystic fibrosis allele, or any other genetic disorder, they may decide that they want children that are free of the disorder. To do this they would use in vitro fertilisation (IVF) to produce embryos. These embryos can be screened, and only those free of the disorder would be implanted.

Advantages of embryo screening	Disadvantages of embryo screening
The children won't have cystic fibrosis.	Embryos with the cystic fibrosis allele are discarded, so we are discarding life; is this ethical?
The cystic fibrosis allele won't be passed on to the next generation.	Some people with the genetic disorder feel that it is discrimination against them.
Saves money in the NHS because there will be fewer sufferers to treat.	

Exam tip AQA

Practise writing out genetic crosses of people with genetic disorders, and predicting the chances of the children having the condition.

Questions

1 What determines whether we are male or female?

2 What causes cystic fibrosis?

3 **H** How does DNA work?

Evidence for previous life forms

Biologists believe that all organisms around today have developed from previous life forms. This theory is called **evolution**. Perhaps the best form of evidence for these earlier life forms is the **fossil** record.

Fossils are the preserved remains of living things from years ago. Fossils form in different ways.

Types of fossils

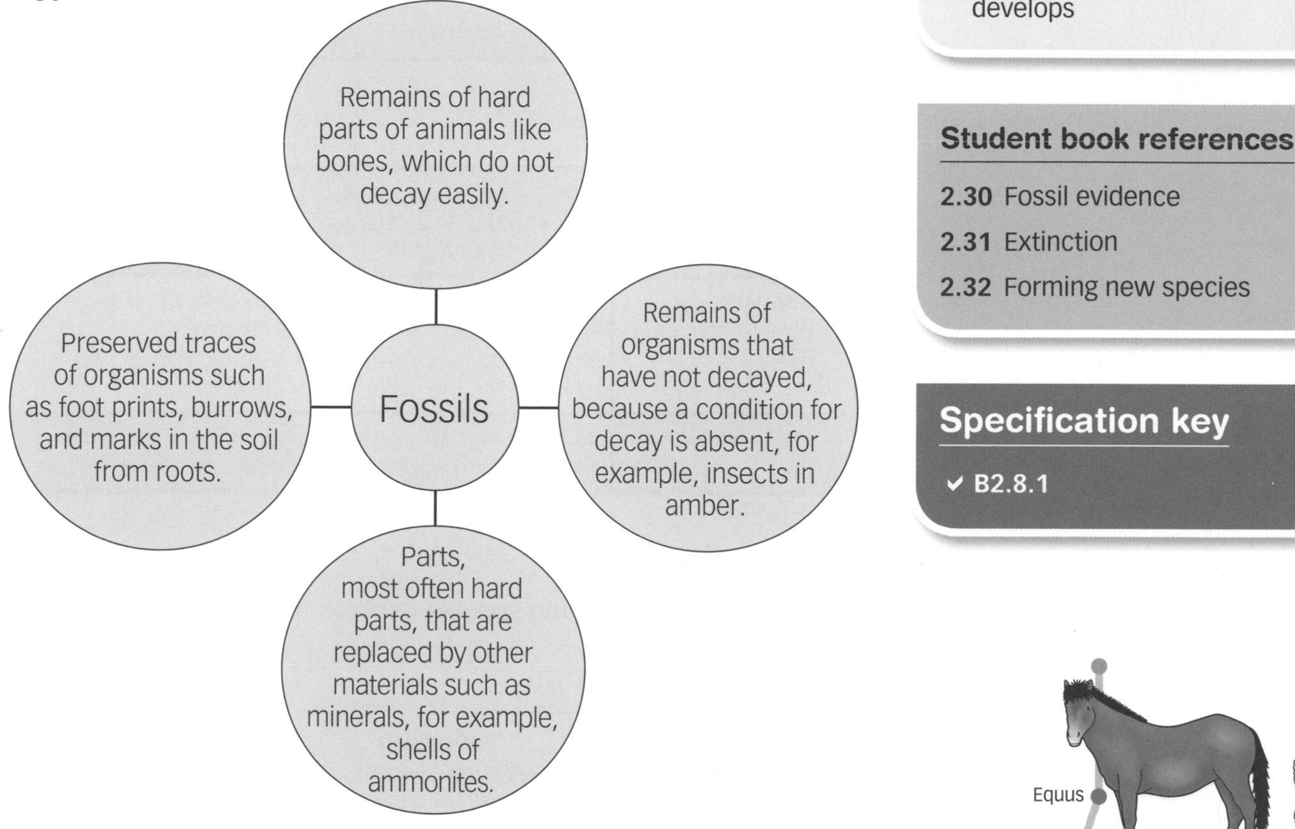

Soft-bodied organisms do not fossilise easily, because the bodies rot. This means we have a poor record of these organisms. They do leave some imprint fossils, which are fine traces in rocks formed from silt. Many of these rocks will have been affected by geological activity, which destroys the fossils. This has resulted in little evidence about the earliest life forms, so scientists are unsure about exactly how life began on Earth.

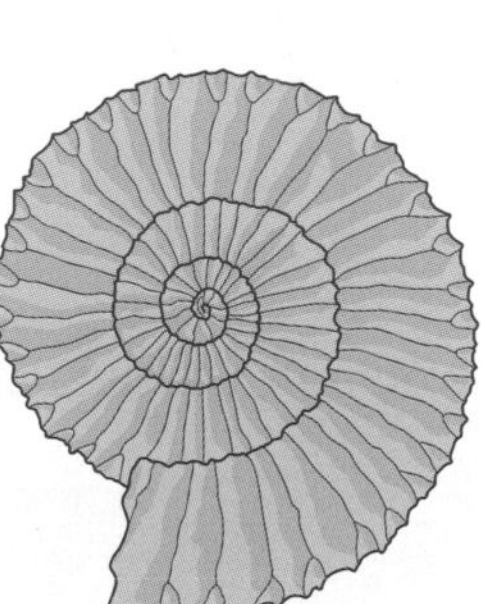

▲ A fossil ammonite.

Fossils can be dated and placed into the correct time sequence. This fossil record shows us how much or how little organisms have gradually changed over time. They can even show us how one species might have changed into another, which is clearly seen in the evolution of the horse.

Revision objectives

- understand how fossils might be formed, and how useful fossils are to scientists
- know what extinction is and how it is caused
- understand the mechanism by which a new species develops

Student book references

2.30 Fossil evidence

2.31 Extinction

2.32 Forming new species

Specification key

✔ B2.8.1

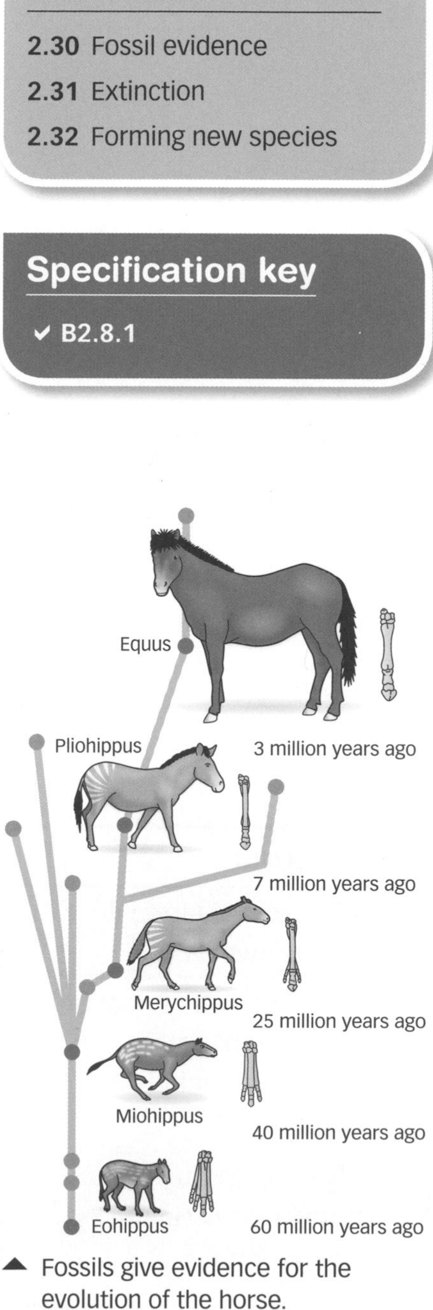

▲ Fossils give evidence for the evolution of the horse.

Key words

evolution, fossil, extinction, speciation, isolation

Exam tip

The fossil record is incomplete. You may be asked to evaluate fossil evidence in a question. Remember that uncertainty arises from the lack of reliable evidence because of an incomplete fossil record for most species.

Extinction

From the fossil record, biologists have found evidence of many examples of organisms that are not alive today. These organisms are extinct.

The causes of extinction

Cause	Effect
1 Changes to the environment over time	Some organisms are not well adapted to cope with the new conditions. For example, as global temperature increased, the woolly mammoths died out.
2 New predators	A more efficient predator hunts a population to extinction, for example, humans hunted dodos.
3 New diseases	Some diseases are so virulent that they destroy a population, for example, the elm tree is almost extinct because of Dutch elm disease.
4 New competitors	A new species may out-compete an existing one, for example, the grey squirrel out-competes the red squirrel, reducing the distribution of the red.
5 A single catastrophic event	Major global events such as asteroid impacts and volcanic eruptions dramatically change the environment. For example, an asteroid impact may have caused the extinction of the dinosaurs.
6 Speciation	As evolution naturally produces new varieties, some will be better adapted, and older versions may die out.

Speciation

Speciation is where one species evolves into two new species.

How new species form

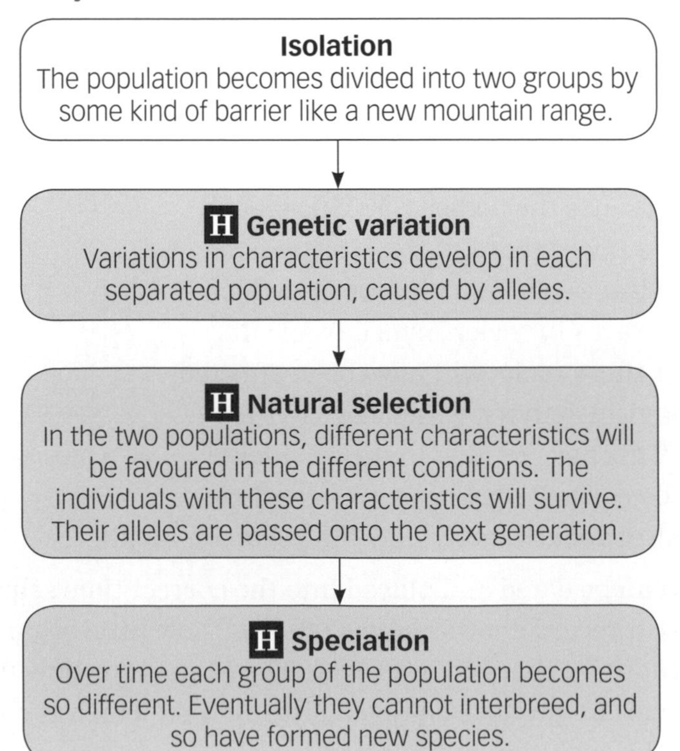

Questions

1 What information do fossils give us?

2 Why are some scientists worried that global warming might lead to extinction?

3 **H** Explain the role of natural selection in speciation.

Questions
Inheritance and evolution

Working to Grade E

1 What is a gene?
2 Who was Mendel?
3 What organisms did Mendel work on?
4 Sex is determined by sex chromosomes.
  a What combination of sex chromosomes does a male have?
  b What combination of sex chromosomes does a female have?
  c Do the gametes of the male or the female contain different sex chromosomes?
5 What do the letters DNA stand for?
6 What is a genetic disorder?
7 Polydactyly is a genetic disorder caused by a dominant allele. What does polydactyly cause?
8 What is a fossil?
9 What is extinction?

Working to Grade C

10 Tongue rolling (T) in humans is dominant to non-tongue rolling (t).
  a In the following cross, show which genes would be in the gametes of the parents.

Parents:	Tongue roller	×	Non-tongue roller
Genes present:	TT	×	tt
Gametes:	◯ ◯		◯ ◯

  b Complete the genetic diagram below to show the outcome of a cross between the two parents.

Gametes		

11 In terms of genetics:
  a What does dominant mean?
  b What does recessive mean?
12 Chromosomes are found in the nucleus of human cells.
  a How many chromosomes are there in a human body cell?
  b How many chromosomes are there in a human sperm cell?
13 Sex is inherited.
  a Use the diagram below to show how sex is inherited.

Parents:	Male	×	Female
Chromosomes present:	____	×	____
Gametes:	◯ ◯		◯ ◯

Gametes		

  b What are the chances that a child born will be a boy?

14 Genes are located on chromosomes. The following diagram shows a pair of chromosomes. The gene for eye colour is marked on one chromosome. Where would it be on the other chromosome?

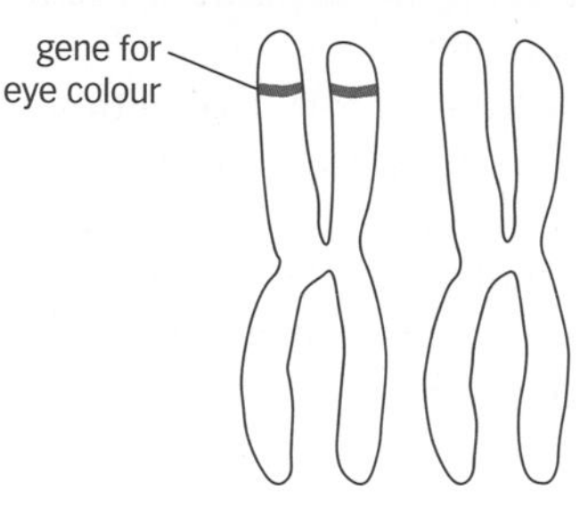

15 Describe the shape of the DNA molecule.
16 What does DNA code for?
17 Polydactyly is a genetic disorder caused by a dominant allele. Look at the family tree below for a family affected by polydactyly.

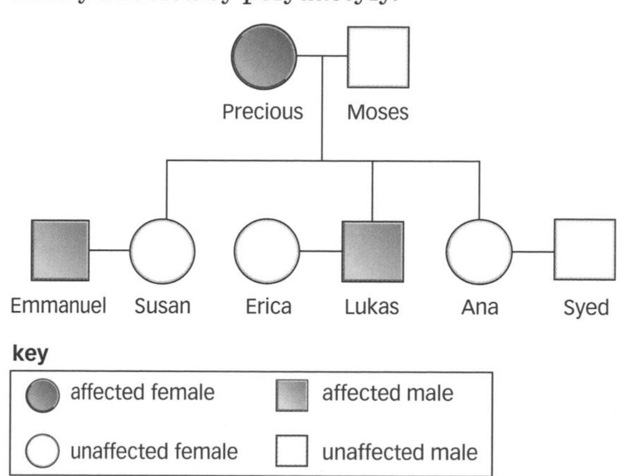

Moses does not have the symptoms. Does he have the polydactyly gene?
18 Are there any humans that might have identical DNA?
19 What uses are made of DNA fingerprinting?
20 Explain why soft-bodied organisms do not fossilise well.
21 Explain why a change in the environment might lead to the extinction of an organism.
22 What is speciation?
23 Why is isolation important for speciation?
24 Explain why scientists are uncertain about the origins of life.
25 How might humans have contributed to the extinction of the dodo?

Working to Grade A*

26 In terms of genetics:
  a what does genotype mean?
  b what does phenotype mean?
  c what does homozygous mean?
27 Brown fur colour (B) in mice is dominant to white (b). Construct a suitable genetic diagram to show the chances of two heterozygous mice producing a white mouse.

28 Mendel did some early work on genetics.

a What were the findings of Mendel's work?

b Why did Mendel struggle to get his work accepted?

c Why do you think it was important that Mendel did not allow his plants to be naturally insect pollinated?

29 Explain why sexual reproduction produces variation in humans.

30 DNA contains four bases.

a How many bases code for each amino acid?

b What is the name given to the section of DNA that codes for one amino acid?

31 Explain how different genes code for different proteins.

32 Look at the family tree for a family affected by polydactyly in question 17, above.

a What is the genotype of Precious and Moses?

b Construct a genetic diagram to show the chances that a fourth child of Precious and Moses would have polydactyly.

33 Embryo screening is sometimes used by couples with a condition like cystic fibrosis in the family.

a What is embryo screening?

b Explain some of the concerns people have against embryo screening.

34 Explain why forensic detectives need to be so careful when collecting samples for DNA fingerprinting.

35 Look at this diagram of the evolution of the elephant family.

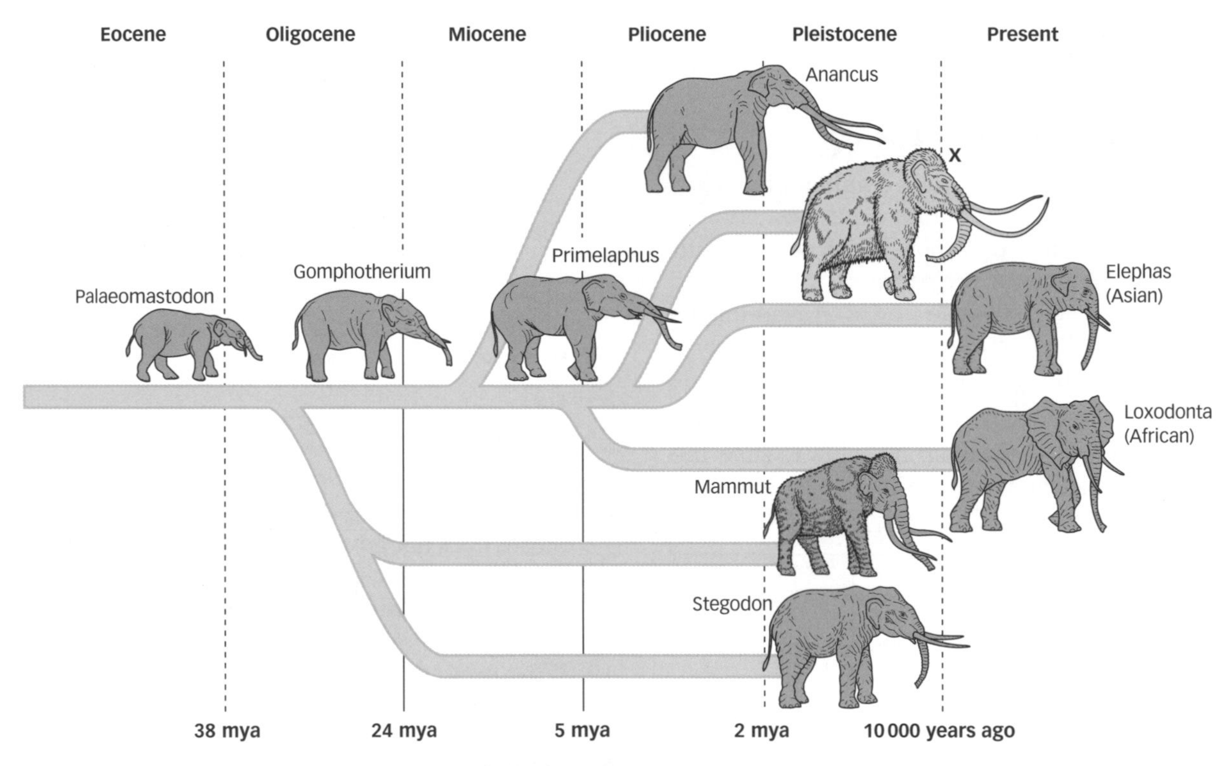

a Many of these animals are now extinct. Biologists know of their existence by studying the fossil record. Which parts of these animals would fossilise best?

b Explain why scientists cannot be absolutely certain about the evolutionary path of some of the elephants.

c What traces might the elephants have left to allow biologists to explain how they walked?

d The animal labelled X is a woolly mammoth. This became extinct many thousands of years ago. What is the most likely cause of its extinction?

36 A biologist called Alfred Russel Wallace studied animal species on either side of the Amazon River. He believed that two different species, one on each side of the river, evolved from the same common ancestor. Explain in detail how these two species might have developed.

1 Red flower colour (R) in geraniums is dominant to white flower colour (r). A plant grower crosses two geranium plants, one red and one white. All of the offspring are red.

a Use the following genetic diagram to show:

i the genes present in the parents *(2 marks)*

ii the genes in the gametes *(2 marks)*

iii the genes present in the offspring. *(1 mark)*

Parents: Red flowers × White flowers

Genes present: ………………… × ……………………

Genes in the gametes: ………………… ……………………

gametes		

(Total marks: 5)

2 The giant panda is an animal that is in danger of becoming extinct.

a What do we mean by the word extinction?

……………………………………………………………………………………………………

……………………………………………………………………………………………………

(1 mark)

b Read the following information about the life of the giant panda.

- The panda is a large land mammal.
- Its diet is composed of mainly bamboo shoots (about 99%).
- There are only about 2000 living in the wild.
- Each adult needs about 9–14 kg of bamboo shoots a day.
- Each bamboo species dies back regularly, and so a panda must have at least two species available as food.
- Pandas have a very low birth rate.
- The panda's habitat is being reduced by humans.

Use the information to suggest **three** reasons why the panda is in danger of extinction.

1 ....

....

2 ....

....

3 ....

....

(3 marks)

(Total marks: 4)

3 The Colorado River created the Grand Canyon in the United States of America. When this happened a species of squirrel evolved into two new species: the Abert and Kaibab tassel-eared squirrels on either side of the canyon.

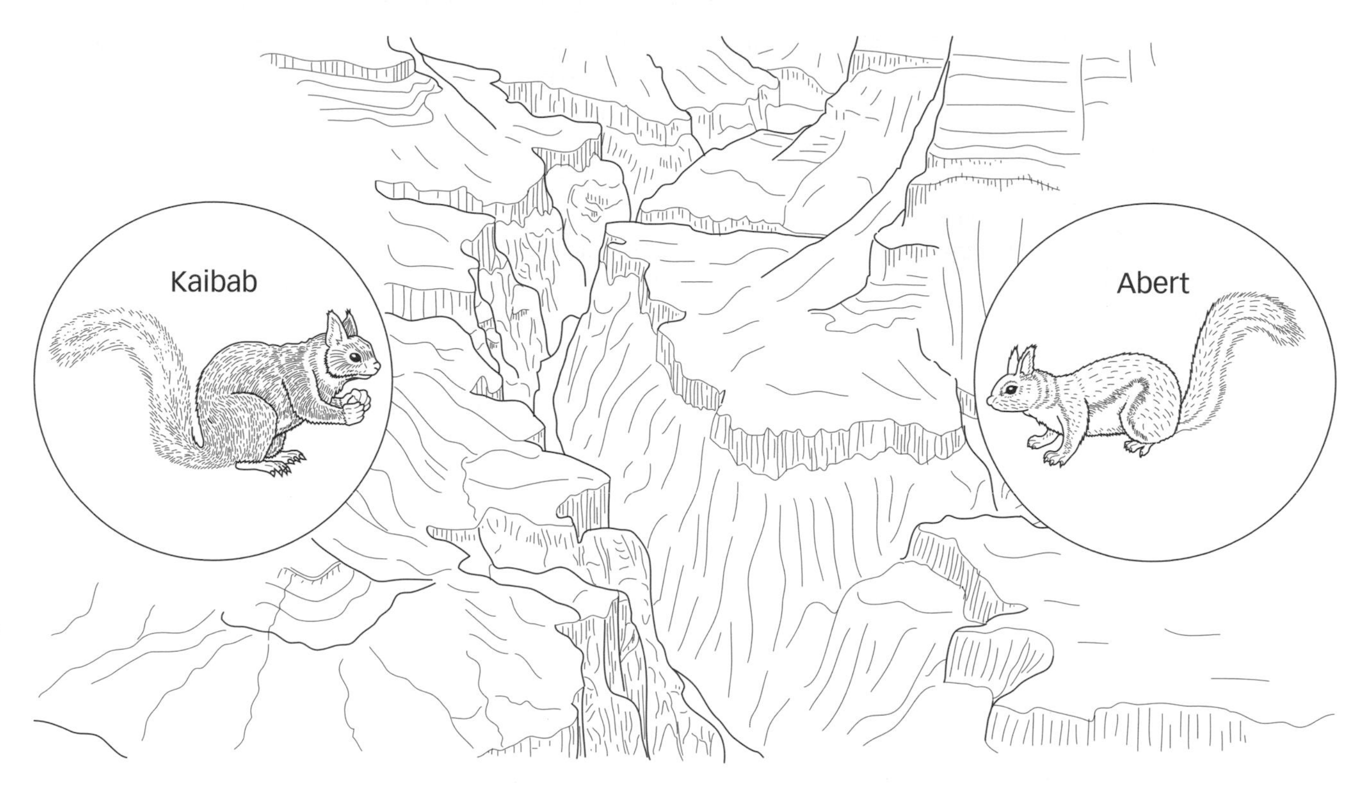

Use your knowledge of speciation to explain how this might have occurred.

....

....

....

....

....

....

....

....

....

....

....

....

....

....

(5 marks)

(Total marks: 5)

How Science Works Genes and proteins, inheritance, gene technology, and speciation

Publish and share!

Mendel is regarded as the father of modern genetics. Today he is thought of as a major figure in the history of biology, but during his lifetime his work was rejected. How can this be? By looking at the way Mendel worked compared to modern scientists we can begin to understand why the importance of his discoveries was overlooked for many years.

Mendel's approach

Mendel set out to investigate the process of inheritance. His approach had four major steps.

The method

Mendel selected varieties of pea plants with clear characteristics. He was careful about controlling the process of pollination, to achieve only the cross he wanted. He avoided contamination by stray pollen from peas not involved in the experiment.

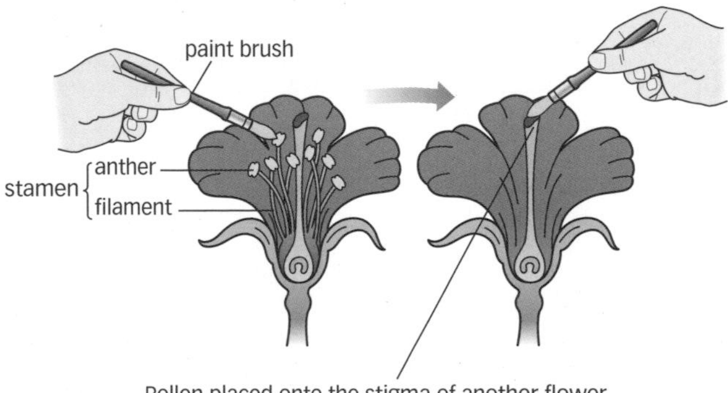

▲ Cross-pollinating flowers.

Mendel kept his experiments fair, by treating all plants in the same way after pollination, and growing all the produced seeds in the same way. He also carried out large numbers of experiments to collect sufficient data to make it reliable.

Data collection

He collected the seeds then grew them. The resulting crop showed different characteristics, and he meticulously counted the numbers of each type that grew.

Analysis of results

When he looked at the results he noticed that certain ratios always occurred, such as the 3:1 ratio. He was able to predict the result of genetic crosses.

Producing conclusions

Using his results he came up with a radical new idea for the process of inheritance. He argued that inheritance was not due to the blending of characteristics, which was the theory at the time. Briefly, he believed that:

- inheritance was controlled by factors
- factors were in pairs
- only one factor was in the gamete
- the offspring got one factor from each parent
- some factors were dominant.

What went wrong?

Failure to spread the word

Mendel tried to share his findings with the scientific community. However, he met with resistance. It was not in line with the ideas of the day, and many rejected the work because of that. Also there were gaps in his theory, as no one knew what these 'factors' were. This made scientists cautious. Mendel failed to get his ideas published widely for the scientific world to look at.

The consequences!

As the work was not published very few scientists knew anything of the work of Mendel. The problem with this was that no one could repeat the experiments to reproduce the results. In addition, other scientists were not aware of the idea of 'factors', so no one looked for them.

How would Mendel have been treated today?

Perhaps the biggest difference today is that scientists are able to publish, even if their results disagree with current views. The key issue is that the methods must be good and fair. Mendel's experimental technique seems sound.

Other scientists check their results. They will seek to reproduce similar results. Those experiments that do give the same results for other scientists will survive. If the experiment cannot be repeated then the ideas are dropped. Mendel's results can be easily repeated.

Today, if we have results that can be reproduced, but not fully explained, like Mendel's, we investigate them further. For example, scientists at the time did not know what factors were. Now scientists all around the world would read about new work, and where there were gaps in the knowledge they would try to explain them. Regarding Mendel's work, about 40 years after his work was written, other scientists found it, and were able to repeat it. Within a few years biologists had discovered the chromosome, and were able to rename factors as genes, which were located on chromosomes.

AQA Upgrade

Answering questions where calculations and interpreting graphs are involved

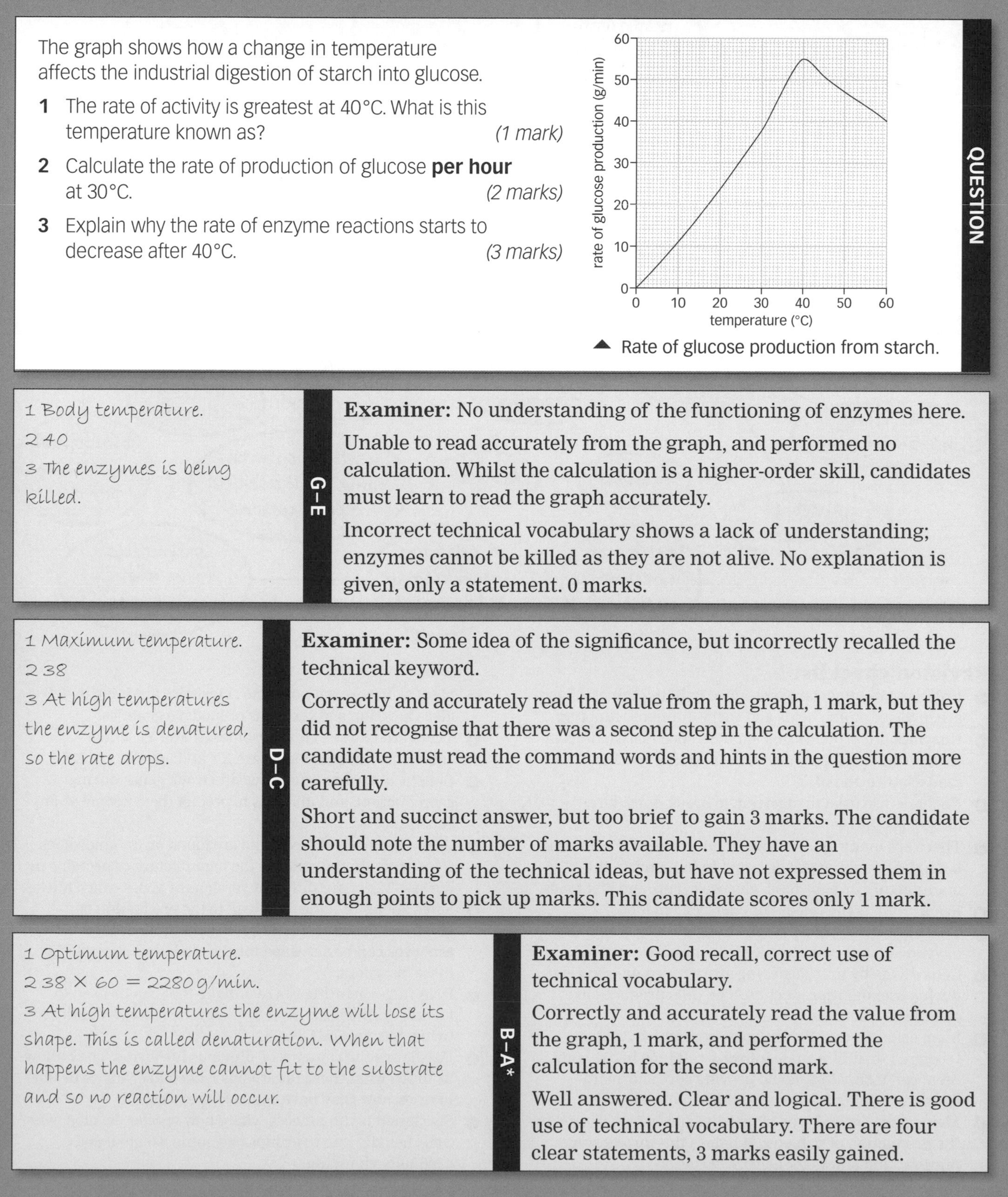

QUESTION

The graph shows how a change in temperature affects the industrial digestion of starch into glucose.

1 The rate of activity is greatest at 40°C. What is this temperature known as? *(1 mark)*

2 Calculate the rate of production of glucose **per hour** at 30°C. *(2 marks)*

3 Explain why the rate of enzyme reactions starts to decrease after 40°C. *(3 marks)*

▲ Rate of glucose production from starch.

G–E

1 Body temperature.
2 40
3 The enzymes is being killed.

Examiner: No understanding of the functioning of enzymes here.

Unable to read accurately from the graph, and performed no calculation. Whilst the calculation is a higher-order skill, candidates must learn to read the graph accurately.

Incorrect technical vocabulary shows a lack of understanding; enzymes cannot be killed as they are not alive. No explanation is given, only a statement. 0 marks.

D–C

1 Maximum temperature.
2 38
3 At high temperatures the enzyme is denatured, so the rate drops.

Examiner: Some idea of the significance, but incorrectly recalled the technical keyword.

Correctly and accurately read the value from the graph, 1 mark, but they did not recognise that there was a second step in the calculation. The candidate must read the command words and hints in the question more carefully.

Short and succinct answer, but too brief to gain 3 marks. The candidate should note the number of marks available. They have an understanding of the technical ideas, but have not expressed them in enough points to pick up marks. This candidate scores only 1 mark.

B–A*

1 Optimum temperature.
2 38 × 60 = 2280 g/min.
3 At high temperatures the enzyme will lose its shape. This is called denaturation. When that happens the enzyme cannot fit to the substrate and so no reaction will occur.

Examiner: Good recall, correct use of technical vocabulary.

Correctly and accurately read the value from the graph, 1 mark, and performed the calculation for the second mark.

Well answered. Clear and logical. There is good use of technical vocabulary. There are four clear statements, 3 marks easily gained.

Part 2 Course catch-up Genes and proteins, inheritance, gene technology, and speciation

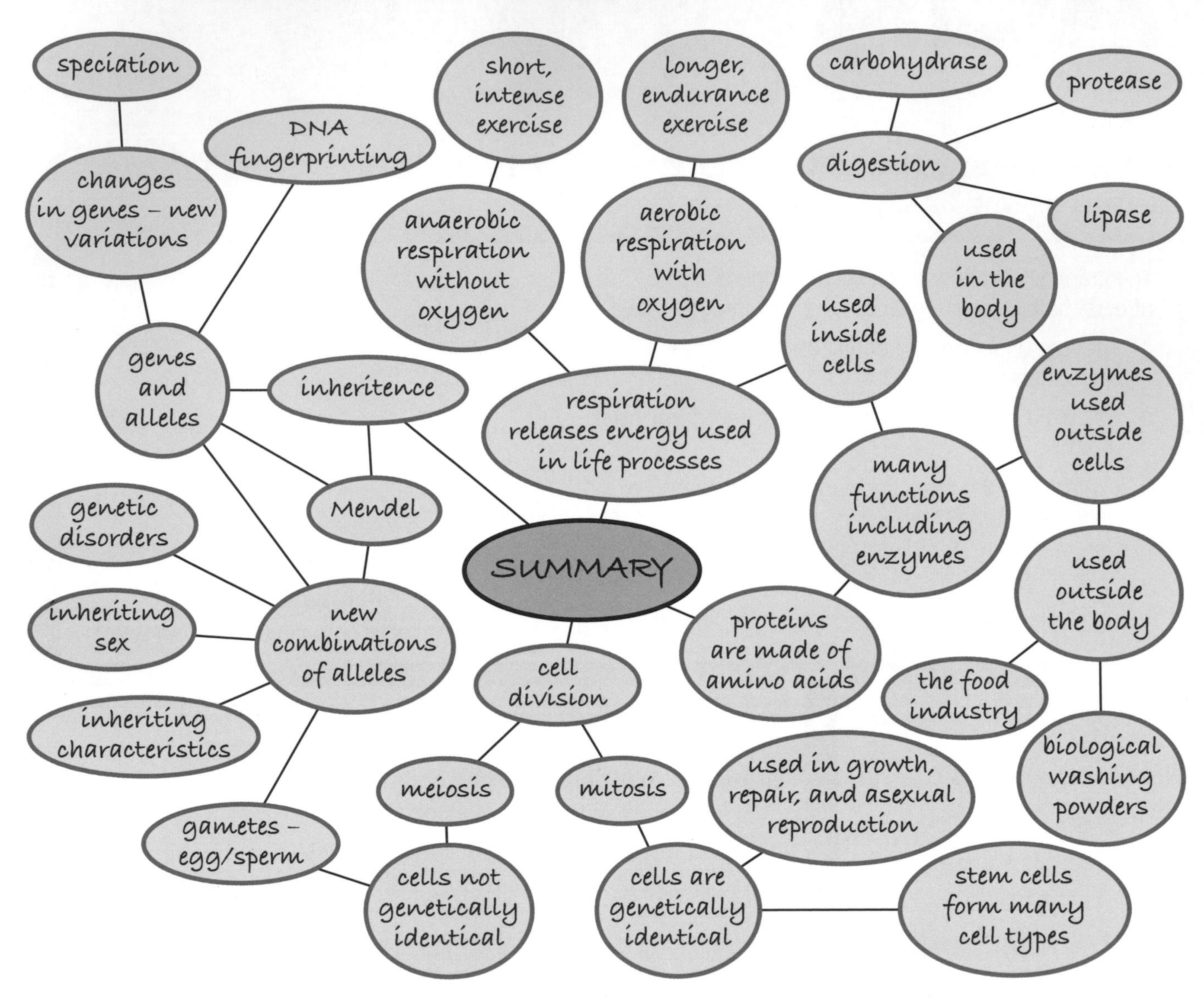

Revision checklist

- Proteins are important molecules in living things. They are made of amino acids and carry out many functions.
- Enzymes are proteins that catalyse specific reactions in living cells. Each enzyme works best at a specific temperature and pH.
- Enzymes are used in digestion to break down large molecules into smaller ones.
- There are many applications of enzymes: in the home in biological washing powders and in industry in the manufacture of baby food, glucose syrup, and diet foods.
- Respiration is the release of energy from foods.
- There are two types of respiration: aerobic, which uses oxygen and anaerobic, which does not use oxygen.
- Cells divide by mitosis during growth, repair, and also during asexual reproduction. The cells produced are genetically identical.
- Stem cells are undifferentiated cells, which can be triggered to differentiate into any cell type in the body. Stem cell technology could provide cures for many diseases, although there are ethical concerns.
- Meiosis is a second type of cell division, which is used in the production of gametes. It halves the chromosome number, and allows for genetic variation.
- Mendel discovered how characteristics are inherited. His ideas have formed the basis of modern genetics.
- There are two sex chromosomes (X and Y), which determine our sex. Males are XY and females are XX.
- Genetic diagrams can be used to track genes during reproduction, and allow us to predict the outcome of any genetic cross.
- Genes are made of DNA, and are found on chromosomes. Genes contain the code for the manufacture of proteins in the cell. There are different versions of genes called alleles.
- Some alleles of genes can lead to the production of abnormal proteins, resulting in genetic disorders. Embryos can be screened for the alleles responsible for these disorders.
- DNA fingerprinting is a technique that can be used to identify individuals because everyone (except identical twins) has unique DNA.
- Fossils are the remains of previous life forms. They allow us to see what these organisms looked like, and to be able to track how they have changed over time.
- Speciation is the process where new species develop over time, usually due to groups becoming isolated, and gradually changing.

Compounds

A **compound** is a substance that is made up of two or more elements. The atoms in a compound are held together by chemical bonds.

Chemical bonding involves the electrons in the highest occupied energy levels, or shells, of atoms. These electrons can be transferred from one atom to another or shared between two atoms.

By transferring or sharing electrons, atoms achieve the stable electronic structure of a noble gas (group 0 elements).

Making ions

When atoms transfer electrons, **ions** are made. An ion is an electrically charged atom, or group of atoms. Electrons are negatively charged, so:

- if an atom loses one or more electrons, it becomes a positive ion
- if an atom gains one or more electrons, it becomes a negative ion.

Most ions have 8 electrons in their highest energy level, like a noble gas atom.

A magnesium atom has 12 protons and 12 electrons. It loses two electrons to become a magnesium ion. The magnesium ion has 12 protons and 10 electrons. This gives it an overall charge of +2. Its formula is Mg^{2+}. You can represent its electronic structure as $[2,8]^{2+}$.

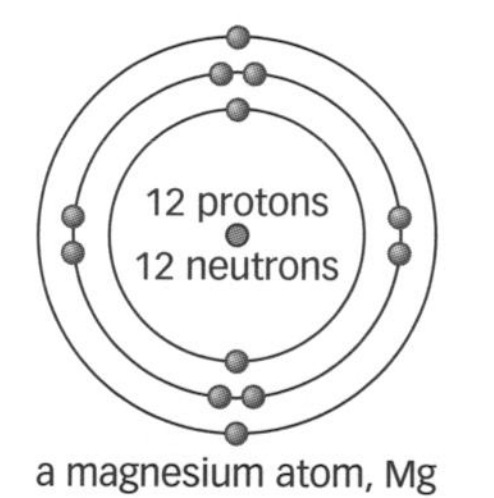

a magnesium atom, Mg

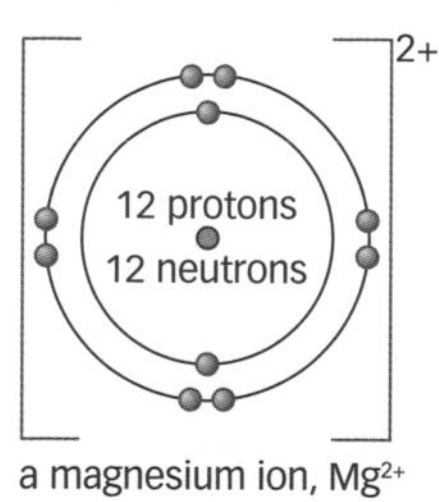

a magnesium ion, Mg^{2+}

An oxygen atom has 8 protons and 8 electrons. It gains two electrons to become an **oxide ion**. The oxide ion has 8 protons and 10 electrons. This gives it an overall charge of –2. Its formula is O^{2-}. Its electronic structure is $[2,8]^{2-}$.

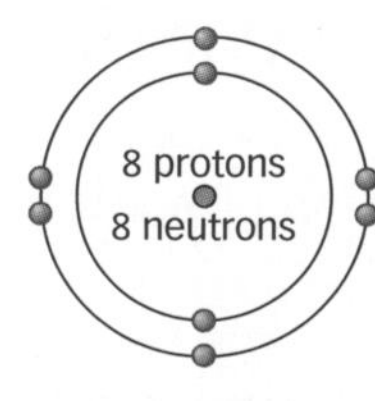

an oxygen atom, O

2–
8 protons
8 neutrons

an oxide ion, O^{2-}

Magnesium oxide is made up of Mg^{2+} ions and O^{2-} ions. Its formula is MgO.

Revision objectives

- ✔ explain how ions form
- ✔ use dot-and-cross diagrams to represent ions
- ✔ write formulae for ionic compounds
- ✔ describe ionic bonding

Student book references

2.3 Ionic bonding
2.4 Making ionic compounds

Specification key

✔ **C2.1.1 a – f**

Key words

compound, ion, oxide ion, alkali metal, halogen, metal halide, ionic compound, giant ionic lattice, ionic bonding

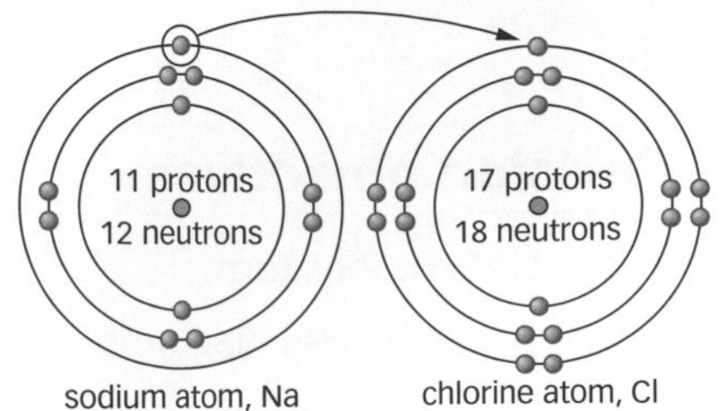

▲ When sodium and chlorine react together, each sodium atom transfers one electron to a chlorine atom.

Reactions of group 1 elements

The elements in group 1 of the periodic table are the **alkali metals**. They react vigorously with group 7 elements, the **halogens**. **Metal halides** are made. For example:

sodium + chlorine → sodium chloride

$$2Na + Cl_2 \rightarrow 2NaCl$$

lithium + bromine → lithium bromide

$$2Li + Br_2 \rightarrow 2LiBr$$

potassium + iodine → potassium iodide

$$2K + I_2 \rightarrow 2KI$$

In these reactions, each metal atom transfers one electron to a halogen atom. This forms:

- metal ions with a single positive charge, such as K^+
- halide ions with a single negative charge, such as I^-.

Potassium iodide is made up of potassium ions, K^+ and iodide ions, I^-. Its formula is KI.

Group 1 metals also react vigorously with oxygen. For example:

lithium + oxygen → lithium oxide

$$4Li + O_2 \rightarrow 2Li_2O$$

The products of the reactions of alkali metals with non-metal elements are all ionic compounds. In each compound the metal ion has a single positive charge.

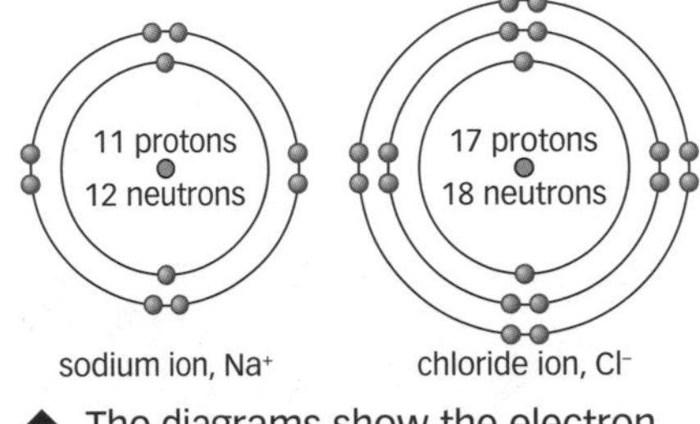

▲ The diagrams show the electron arrangements in sodium and chloride ions.

Inside ionic compounds

Compounds made up of ions are **ionic compounds**. An ionic compound is a giant structure of ions. The ions are held together by strong electrostatic forces of attraction between the oppositely charged ions. These forces act in all directions. They hold the ions in a regular pattern called a **giant ionic lattice**. This is **ionic bonding**.

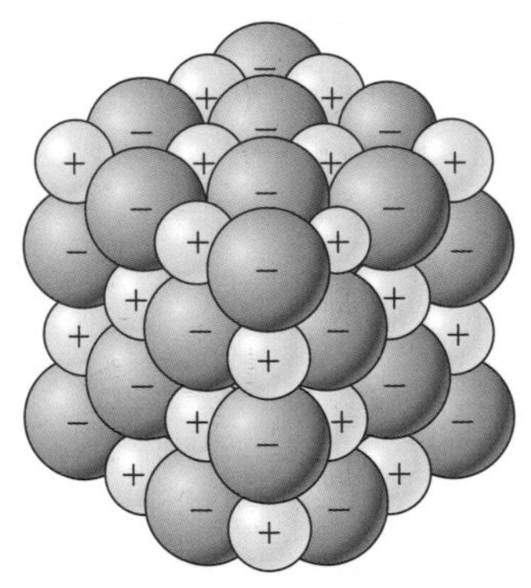

▲ The giant ionic lattice structure of sodium chloride.

Exam tip

AQA

Remember, nearly all ions have 8 electrons in their highest energy level.

Questions

1. Draw dot-and-cross diagrams to represent the electronic structures of the ions in sodium chloride and calcium chloride.
2. Write a word equation for the reaction of sodium with bromine.
3. Describe the bonding in a crystal of sodium chloride.

Covalent bonds

In some elements and compounds, atoms share pairs of electrons. They do this to achieve the stable electronic structure of a noble gas. A shared pair of electrons is called a **covalent bond**. Covalent bonds are very strong.

Simple molecules

Some covalently bonded substances consist of simple molecules. A simple molecule is made up of a small number of atoms, with covalent bonds between the atoms. For example:

- Chlorine exists as chlorine molecules, Cl_2. A chlorine molecule consists of two chlorine atoms. Each chlorine atom shares one of its electrons with the other chlorine atom to make a covalent bond.

In this way, each chlorine atom has a share of eight electrons in its highest occupied energy level. The chlorine atoms now have the stable electronic structure of argon, a noble gas.

- Hydrogen exists as hydrogen molecules, H_2. Each hydrogen atom has the stable electronic structure of the noble gas helium.
- An oxygen molecule, O_2, consists of two oxygen atoms. Two pairs of electrons are shared between the atoms. The two shared pairs of electrons form a strong **double covalent bond**.

The atoms in the compounds in the table are joined together by covalent bonds to make simple molecules. Each line in a displayed formula represents a covalent bond.

Name of compound	Molecular formula	Dot-and-cross diagrams	Displayed formula
hydrogen chloride	HCl	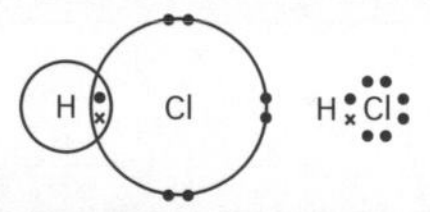	H—Cl
water	H_2O	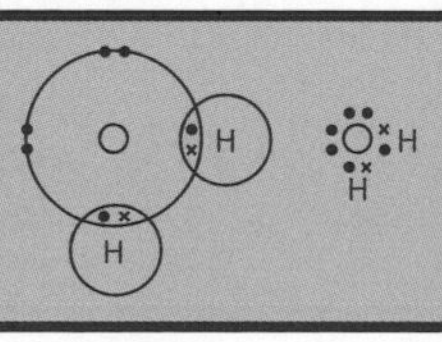	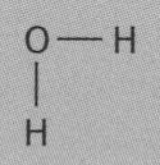 O—H, O—H (displayed structure)
ammonia	NH_3	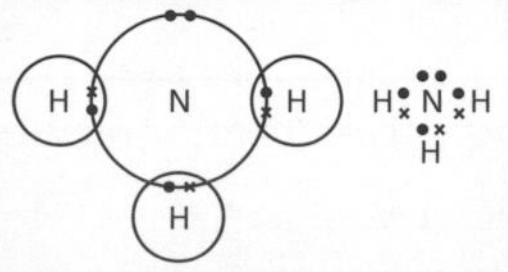	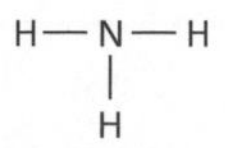 H—N—H, N—H
methane	CH_4	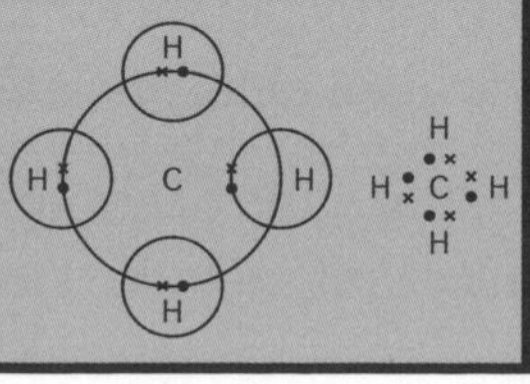	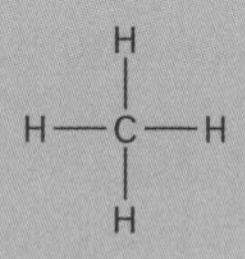 H—C—H with H above and below C

Revision objectives

- explain how covalent bonds are formed
- draw dot-and-cross diagrams for simple molecules
- describe metallic bonding

Student book references

2.1 Covalent bonding
2.2 More about molecules
2.5 Inside metals

Specification key

✔ C2.1.1 g – i

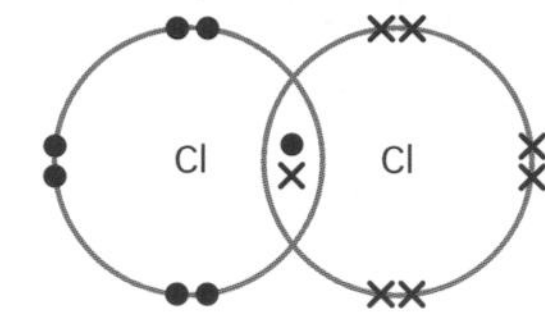

▲ This dot-and-cross diagram shows the electrons in the highest occupied energy levels of a chlorine molecule, Cl_2.

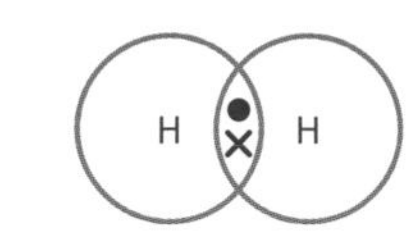

▲ This dot-and-cross diagram shows the electrons in a hydrogen molecule, H_2.

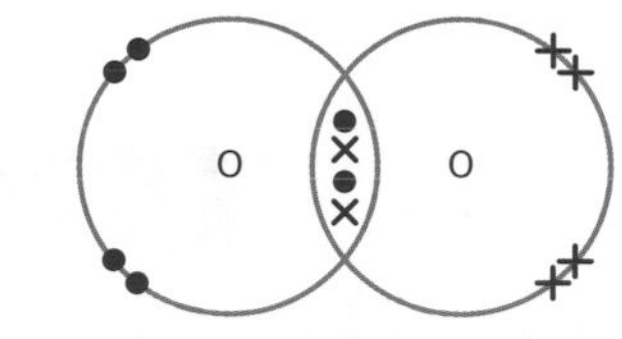

▲ Two pairs of electrons are shared between the atoms in an oxygen molecule.

Giant structures

The atoms in some covalently bonded substances are joined together in huge networks called **giant covalent structures**, or **macromolecules**.

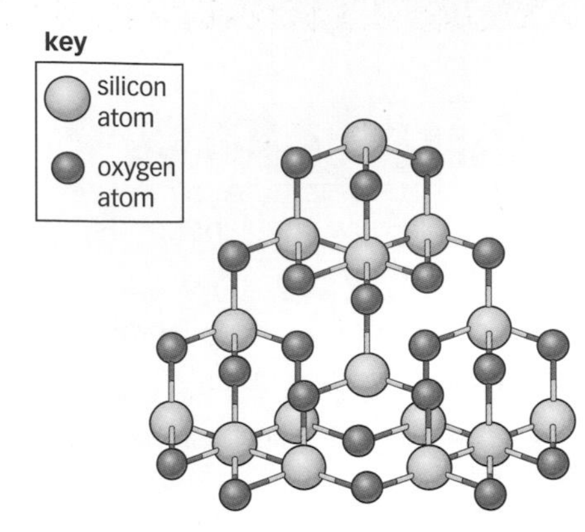

The atoms in silicon dioxide are also joined together in a giant covalent structure.

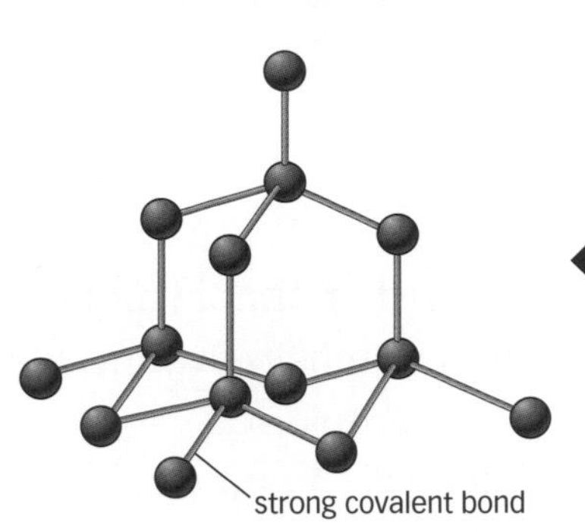

Diamond is a form of the element carbon. In diamond, strong covalent bonds join each carbon atom to four other carbon atoms. The diagram shows just a tiny part of the structure of diamond.

Metallic bonding

Metals consist of giant structures of atoms arranged in a regular pattern.

H In metals, the electrons in the highest energy levels of the atoms become **delocalised**. These electrons are no longer part of single atoms, but are free to move through the whole structure of the metal. The atoms that have lost electrons are positive ions. These are arranged in a regular pattern as a **giant metallic structure**. There are strong electrostatic forces of attraction between the positive ions and the moving delocalised electrons. This is **metallic bonding**.

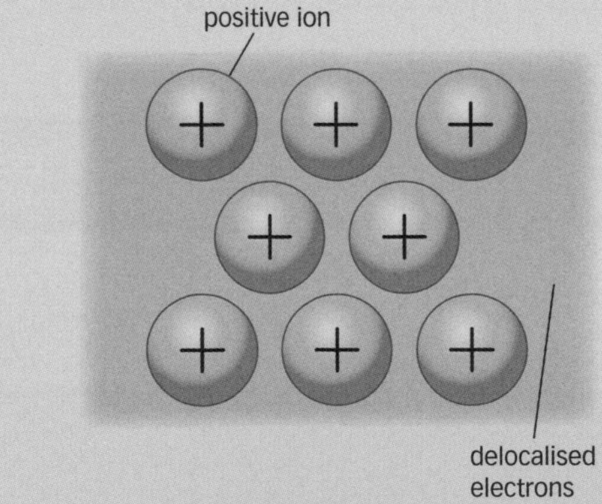

Metallic bonding.

Key words

covalent bond, simple molecule, double covalent bond, giant covalent structure, macromolecule, delocalised, giant metallic structure, metallic bonding

Exam tip AQA

Remember, there are covalent bonds in compounds of non-metals and in most non-metal elements.

Questions

1 Describe the differences between simple molecules and macromolecules.

2 Draw displayed formulae for molecules of hydrogen, oxygen, hydrogen chloride, and water.

3 **H** Describe the differences between the bonding in giant covalent and giant metallic structures.

Working to Grade E

1 Tick the boxes to show which substances listed below are compounds.
   - a Oxygen ☐
   - b Carbon dioxide ☐
   - c Magnesium chloride ☐
   - d Chlorine ☐
   - e Germanium ☐
   - f Potassium bromide ☐
   - g Water ☐

2 Highlight the statements below that are true. Then write corrected versions of the statements that are false.
   - a A compound is made up of atoms of two or more elements.
   - b When atoms share electrons, they form ionic bonds.
   - c Diamond has a giant covalent structure.
   - d Elements in group 1 form ions with a charge of +1.
   - e Elements in group 7 form ions with a charge of +7.

3 Choose words from the box below to fill in the gaps in the sentences that follow. The words in the box may be used once, more than once, or not at all.

positively	noble	gas
molecules	metals	ions
negatively	alkali	seven
zero	one	eight

When atoms form chemical bonds by transferring electrons, they form ________. Atoms that lose electrons become ________ charged ________. Atoms that gain electrons become ________ charged ions. Ions have the electronic structure of a ________ ________, or group ________ atom.

4 Name the compounds formed when the following pairs of elements react together.
   - a Sodium and chlorine.
   - b Lithium and oxygen.
   - c Potassium and iodine.
   - d Sodium and bromine.

5 Draw lines to link each diagram to the type of structure that it represents.

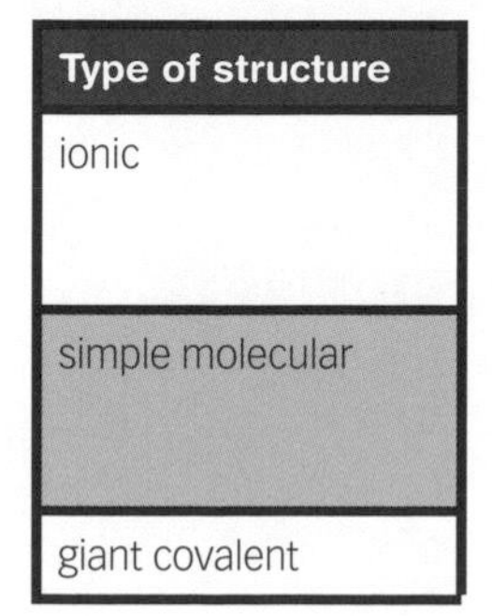

Type of structure
ionic
simple molecular
giant covalent

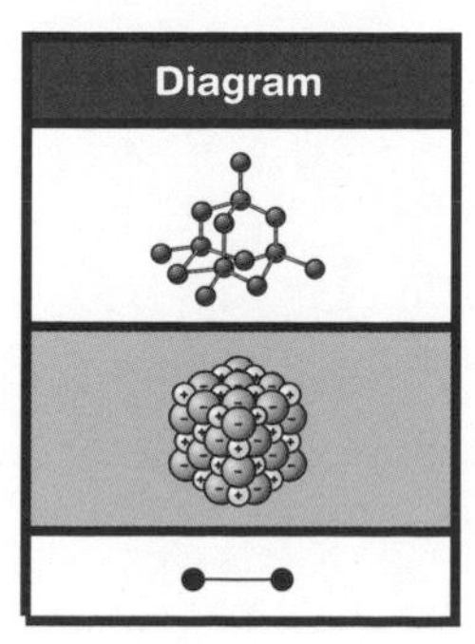

Working to Grade C

6 Tick the boxes to show which formulae below represent compounds.
   - a Cl_2 ☐
   - b Na ☐
   - c HCl ☐
   - d H_2 ☐
   - e CH_4 ☐
   - f O_2 ☐

7 Complete the table to show the numbers of protons and electrons in the ions listed. Use data from the periodic table to fill in the first empty column.

Ion	Number of protons	Number of electrons
Li^+		
F^-		
Na^+		
Cl^-		
Mg^{2+}		
Br^-		
Ca^{2+}		

8 Write the electronic structures of the ions below.
   - a Na^+
   - b Cl^-
   - c Mg^{2+}
   - d O^{2-}
   - e Ca^{2+}
   - f F^-

9 Give the formulae of the compounds that are made up of the ions listed in the table.

Formula of positive ion	Formula of negative ion	Formula of compound
Na^+	Cl^-	
Mg^{2+}	O^{2-}	
Ca^{2+}	Cl^-	
Rb^+	O^{2-}	

10 Draw dot-and-cross diagrams and displayed formulae to show the atoms and covalent bonds in the substances listed below.
   - a H_2
   - b Cl_2
   - c O_2
   - d HCl
   - e H_2O
   - f NH_3
   - g CH_4

Working to Grade A*

11 The diagram below represents the bonding in a metal. Use the diagram to help you explain the structure of the metal, and how it is held together.

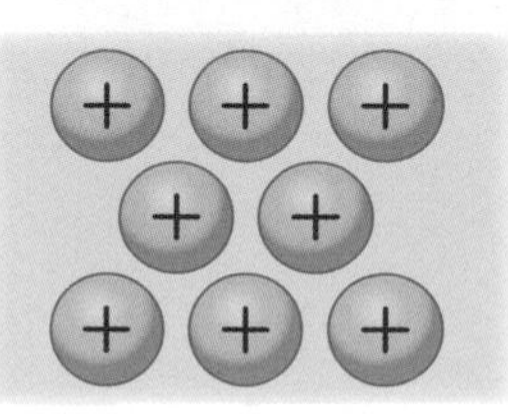

Examination questions
Structure and bonding

1 a Potassium chloride is an ionic compound.

i Complete the sentences below.

Potassium chloride is made up of two types of particle. These particles are positively charged potassium ions and .... charged .... ions.

(2 marks)

ii Use the periodic table to help you work out the number of protons in a potassium ion.

....

(1 mark)

iii Complete the diagram below to show all the electrons in a potassium ion.

....

(2 marks)

iv Name and describe the forces that hold solid potassium chloride together.

....

....

....

(3 marks)

1 b A chemist makes potassium chloride in the laboratory from two starting materials: potassium and chlorine.

i Name the type of bonding in the element potassium.

....

(1 mark)

ii Name the type of bonding in a chlorine molecule.

....

(1 mark)

iii The diagram below shows the electrons in the highest occupied energy level (outer shell) of a chlorine atom.

Add to the diagram to show how the electrons are arranged in a chlorine molecule. Show the electrons in the highest occupied energy levels (outer shells) **only**.

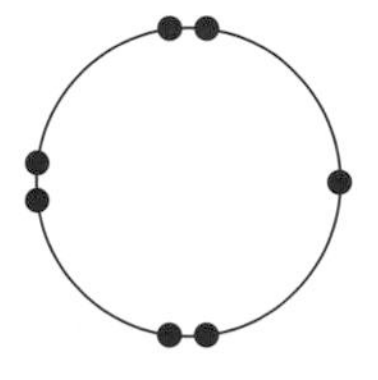

(2 marks)

(Total marks: 12)

3: Inside molecules, metals, and ionic compounds

Revision objectives

- explain the links between structure and properties for simple molecular substances, ionic compounds, and metals

Student book references

2.6 Molecules and properties
2.7 Properties of ionic compounds
2.10 Metals

Specification key

✔ C2.2.1 ✔ C2.2.2 ✔ C2.2.4

Molecules

Many non-metal elements, and compounds made up of non-metals, consist of simple molecules. These substances:

- do not conduct electricity, because the molecules do not have an overall electric charge
- have relatively low melting points and boiling points.

H In substances that consist of simple molecules:

- the covalent bonds that join the atoms together in a molecule are strong
- the forces of attraction between each molecule and its neighbours – the **intermolecular forces** – are weak.

It is the weak intermolecular forces that must be overcome when a substance melts or boils, not the covalent bonds. This is why substances that consist of molecules have low melting and boiling points.

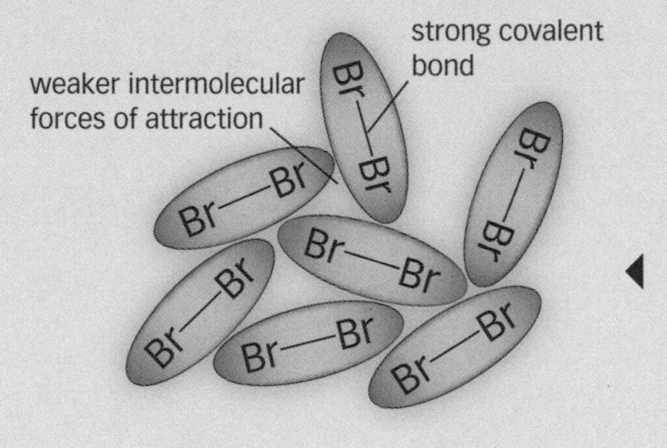

This diagram shows the covalent bonds and intermolecular forces in liquid bromine.

Ionic compounds

Sodium chloride is an ionic compound. It is made up of positive sodium ions and negative chloride ions. These ions are arranged in a regular pattern called a **giant ionic lattice**.

There are very strong electrostatic forces of attraction between the sodium and chloride ions. The forces act in all directions.

Much energy is needed to break the strong bonds between the ions in sodium chloride, so sodium chloride has a high melting point and a high boiling point.

All ionic compounds have high melting points and boiling points because of the large amounts of energy needed to break their many strong bonds.

Solid ionic compounds do not conduct electricity. This is because their ions are not free to move from place to place to carry the current.

If you melt an ionic compound, or dissolve it in water, its ions become free to move. The free ions carry the current.

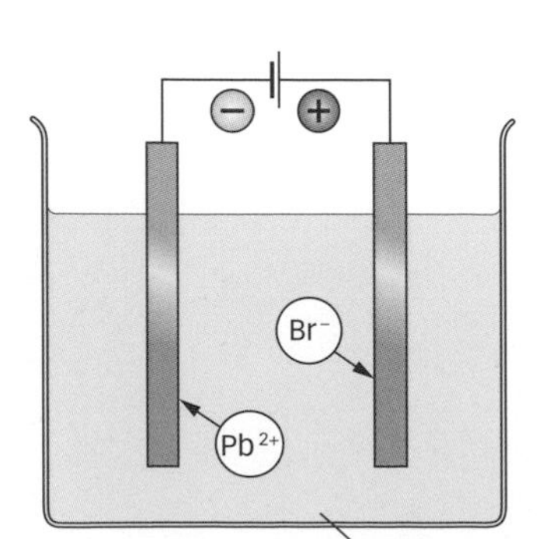

Lead bromide is an ionic compound. Liquid lead bromide conducts electricity because its ions are free to move towards the electrodes.

Metals

H Metals conduct heat and electricity. This is because their delocalised electrons are free to move throughout the metal.

In metals, the layers of atoms can slide over each other easily. This is why it is easy to bend metals, and make them into different shapes.

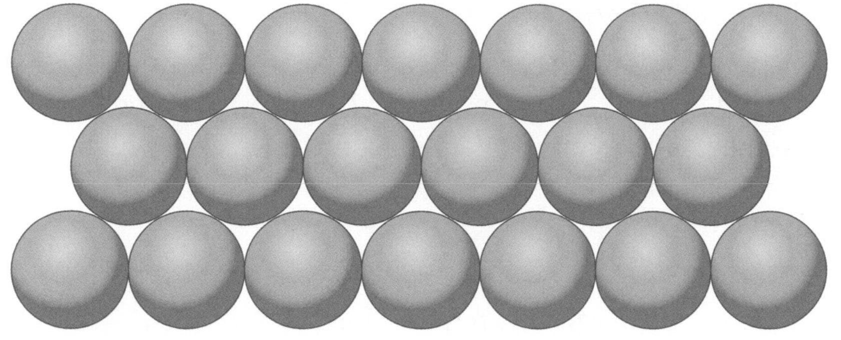

▲ The atoms in a metal.

An **alloy** is a mixture of a metal with one or more other elements. The other elements are usually metals.

The different elements in an alloy have atoms of different sizes. The different sized atoms distort the layers in the metal structure. This makes it more difficult for the layers to slide over each other. So alloys are harder than pure metals.

Shape-memory alloys such as Nitinol are used to make dental braces. They can go back to their original shape after being bent or twisted.

Key words

intermolecular force, giant ionic lattice, alloy, shape-memory alloy

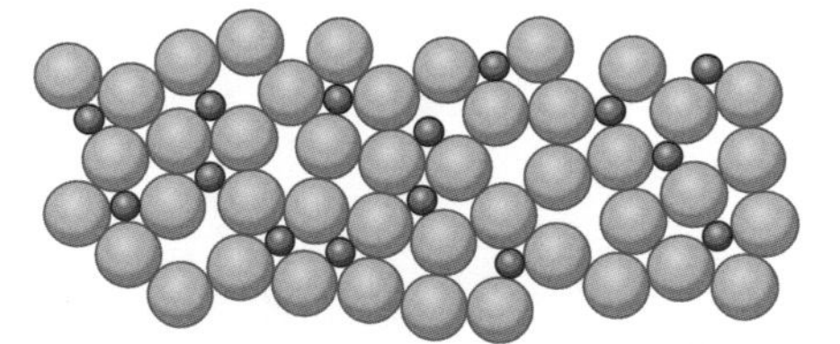

▲ An alloy is a mixture of a metal with small amounts of one or more other elements.

Questions

1. Explain why substances that consist of simple molecules do not conduct electricity.
2. Draw a table to compare the properties of ionic compounds, metals, and substances that consist of simple molecules.
3. **H** Explain why substances that consist of simple molecules have relatively low melting and boiling points.

Exam tip

AQA

You may be given the properties of a substance, and asked to suggest what type of structure it has. Practise doing this before the exam.

4: The structures of carbon, and nanoscience

Revision objectives

- recognise the structures of diamond and graphite, describe their properties and explain their uses
- describe some applications of nanoscience

Student book references

2.8 Diamond and graphite
2.9 Bucky balls and nanotubes

Specification key

✔ C2.2.3 ✔ C2.2.6

Giant covalent structures

Some covalently bonded substances form giant structures, or macromolecules. For example:

- **Diamond** – a form of carbon.
- **Graphite** – another form of carbon.
- Silicon dioxide.

The atoms in these substances are arranged in huge repeating patterns, or **lattices**.

There are very strong covalent bonds between the atoms. Large amounts of energy are needed to break these bonds, so substances with giant covalent structures have very high melting points.

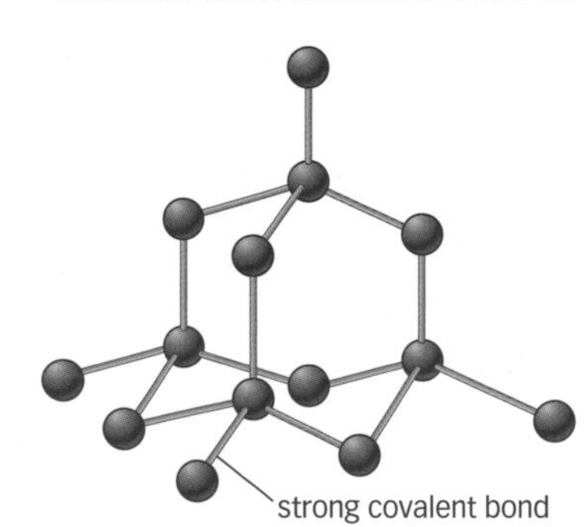

▲ Part of the structure of diamond.

Diamond

In diamond, each carbon atom is joined to four other carbon atoms by strong covalent bonds. A giant covalent structure results. This explains why diamond is very hard.

Graphite

The carbon atoms in graphite have a different arrangement to those in diamond. This gives graphite and diamond different properties.

In graphite, each atom forms covalent bonds with three other atoms, making a layer. A lump of graphite consists of many of these layers. There are no covalent bonds between the layers, just weak intermolecular forces. This means the layers can slide over each other, making graphite soft and slippery.

H For every carbon atom of graphite, three of its four electrons in the highest energy level are involved in covalent bonding. The other electron in each carbon atom is delocalised, or free to move. The delocalised electrons in graphite can carry an electric current and help conduct heat.

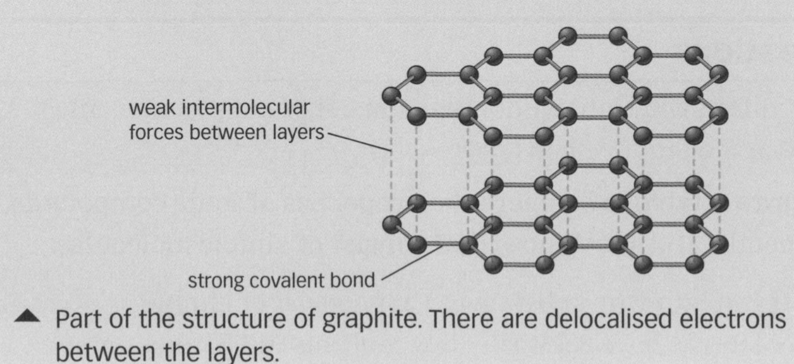

▲ Part of the structure of graphite. There are delocalised electrons between the layers.

H Fullerenes

Carbon can also exist as **fullerenes**. Fullerenes are a type of carbon made up of hexagonal rings of carbon atoms.

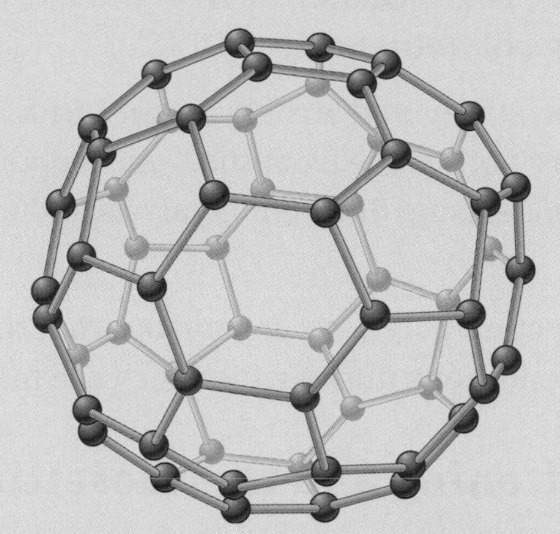

▲ One common fullerene is **Buckminsterfullerene**. It consists of molecules made up of 60 carbon atoms, arranged to form a hollow sphere.

Fullerenes are hollow. The space inside is big enough for atoms and small molecules to fit in. This means that fullerenes can be used to deliver drugs to specific places in the body. Fullerenes are also useful as lubricants, and as catalysts.

Fullerene molecules join up to make **nanotubes**. Nanotubes are used to reinforce materials, for example, in graphite tennis racquets.

Nanoscience

Nanoparticles are particles that are made up of a few hundred atoms. They have diameters of between 1 nanometre (nm) and 100 nm. **Nanoscience** is the study of nanoparticles.

Nanoparticles have different properties to those of the same substance in normal-sized pieces. Their surface area is high compared to their volume.

The special properties of nanoparticles may lead to new developments, including:

- new computers
- new catalysts
- new waterproof coatings
- stronger and lighter construction materials
- new cosmetics such as sun-protection creams and deodorants.

Key words

diamond, graphite, lattice, fullerene, Buckminsterfullerene, nanotube, nanoparticle, nanoscience

Exam tip AQA

You need to know some applications of nanoscience, but you do not need to remember specific examples or properties.

Questions

1 Describe the bonding in diamond.

2 Explain why diamond and graphite have different properties.

3 List **five** ways in which nanoparticles may be used in the future.

Revision objectives

- explain how the properties of polymers are linked to what they are made from, and the conditions under which they are made
- explain how the uses of polymers are linked to their structures

Student book references

2.11 Explaining polymer properties

Specification key

- C2.2.5

Key words

high-density poly(ethene), low-density poly(ethene), thermosoftening polymer, thermosetting polymer

Exam tip

AQA

Practise explaining the difference in properties between thermosoftening and thermosetting polymers.

Question

1 Make a table to summarise the differences in properties and structure between thermosoftening and thermosetting polymers.

Two types of poly(ethene)

There are two types of poly(ethene):

- **high-density poly(ethene)**, or HDPE
- **low-density poly(ethene)**, or LDPE.

HDPE is denser, stiffer, and stronger than LDPE. The two types of polythene are both made from the same monomer, ethene. But they are made using different catalysts and under different conditions.

The properties of all polymers depend on what they are made from and the conditions under which they are made.

Thermosoftening and thermosetting polymers

You can divide polymers into two groups:

- **Thermosoftening polymers** – soften easily when warmed, and can easily be moulded into new shapes. They can be recycled.
- **Thermosetting polymers** – do not melt when they are heated. They cannot be recycled.

The structures of the two types of polymer explain their properties. Thermosoftening polymers consist of individual, tangled polymer chains.

H The forces of attraction between the separate chains are weak.

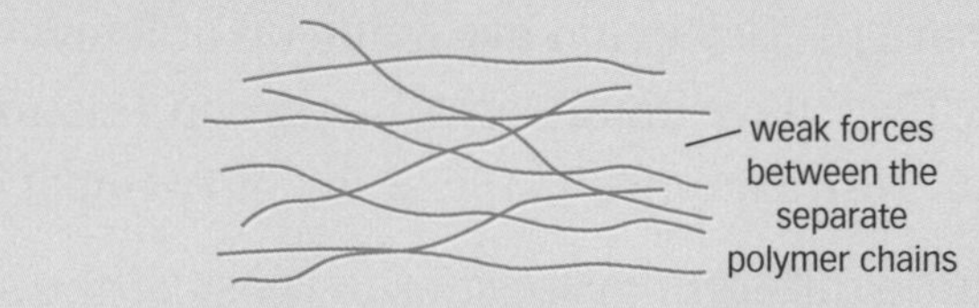

▲ Polymer chains in a thermosoftening polymer.

Thermosetting polymers consists of polymer chains with cross-links between them. The cross-links prevent them from melting.

H The cross-links are strong intermolecular bonds.

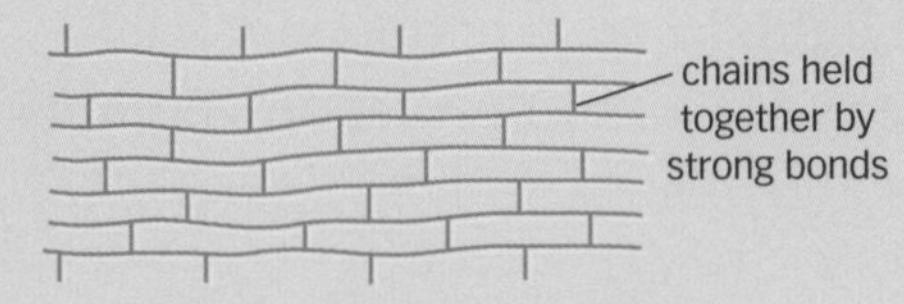

▲ Polymer chains in a thermosetting polymer.

Questions
Structure, properties, and uses

Working to Grade E

1 Highlight the correct word or phrase in each pair of **bold** words in the sentences that follow.

Methane consists of simple molecules.
It has a **high/low** melting point. It **does/does not** conduct electricity because its molecules **have/do not have** an overall electric charge.

2 Highlight the statements below that are true. Then write corrected versions of the statements that are false.

a The forces between the oppositely charged ions in ionic compounds are weak.

b Ionic compounds have high boiling points.

c Ionic compounds conduct electricity when solid.

d Ionic compounds do not conduct electricity when dissolved in water.

e When ionic compounds conduct electricity, the ions carry the current.

3 Choose words from the box below to fill in the gaps in the sentences that follow. The words in the box may be used once, more than once, or not at all.

softer	stiff
molecules	atoms
bendy	more
harder	less

The layers of ______ in metals can slide over each other easily. This makes metals ______. Alloys are ______ than pure metals because the different-sized atoms in the structure make it ________ easy for the layers to slide over each other.

4 **a** Give the name of **one** shape-memory alloy.

b Give **one** use of this shape-memory alloy.

Working to Grade C

5 Use the data in the table to answer the questions below it.

Substance	Does the solid conduct?	Does the liquid conduct?	Melting point (°C)	Boiling point (°C)
J	No	Yes	801	1413
K	No	No	–182	–162
L	Yes	Yes	1083	2595
M	No	No	> 3550	4837
N	No	Yes	2852	3600

a Identify **two** ionic compounds in the table.

b Which letter represents copper?

c Which letter represents diamond?

d Predict the hardest substance in the table.

e Identify the **two** substances with covalent bonds.

f Predict the substance that can easily be bent.

6 Explain why:

a thermosoftening polymers can be recycled

b thermosetting polymers cannot be recycled

c low-density poly(ethene) and high-density poly(ethene) have different properties.

7 The diagrams below show the structures of graphite and diamond. Use the diagrams to explain why diamond is hard and why graphite is soft and slippery.

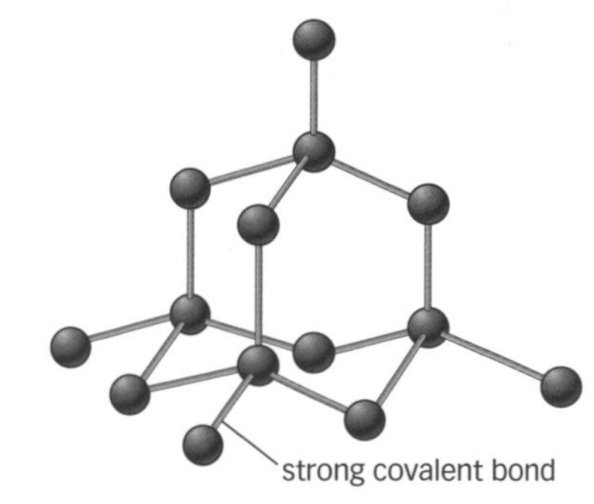

▲ Diamond.

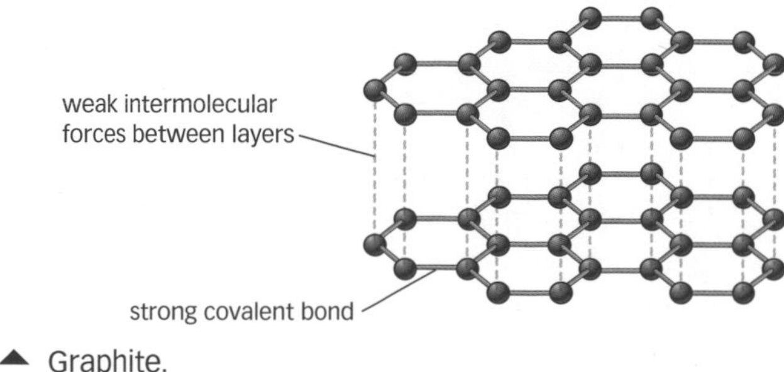

▲ Graphite.

Working to Grade A*

8 **a** Complete the sentences below.
Fullerenes are a form of the element ________.
The structure of fullerenes is based on ________ rings of atoms of this element.

b List **four** uses of fullerenes.

9 Write an **M** next to the sentences below that are true of metals only. Write a **G** next to the sentences that are true of graphite only. Write a **B** next to the sentences that are true of both metals and graphite.

a The structure includes delocalised electrons. ___

b There are strong covalent bonds between the atoms. ___

c The structure is arranged in layers. ___

d The structure includes positive ions. ___

e The layers can slide over each other. ___

10 The table shows the melting and boiling points of two substances whose atoms are joined together by covalent bonds.
Use ideas about bonding and intermolecular forces to explain the differences in melting and boiling points.

Substance	Melting point (°C)	Boiling point (°C)
silicon dioxide	1610	2230
nitrogen dioxide	−11	21

11 The table gives data about two different forms of carbon: graphite and diamond.
Use ideas about delocalised electrons to explain the difference shown.

Substance	Does it conduct electricity?
diamond	no
graphite	yes

12 Use ideas about intermolecular forces to explain why thermosoftening polymers melt at low temperatures, and why thermosetting polymers do not melt when they are heated.

Examination questions
Structure, properties, and uses

1 a The table below gives data about the properties of low-density poly(ethene) (LDPE) and high-density poly(ethene) (HDPE).

	LDPE	HDPE
density (g/cm^3)	0.92	0.95
strength (MPa)	12	31
transparency	good transparency	less transparent
relative flexibility	flexible	stiff

Draw a (ring) around the correct answer in each box to complete the sentence.

HDPE is more suitable for making garden furniture than LDPE

because HDPE is [weaker / stronger / less transparent] and [stiffer. / more flexible. / less dense.]

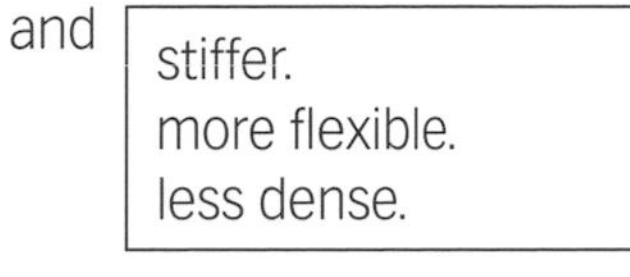

(2 marks)

b The diagram shows part of a poly(ethene) molecule.

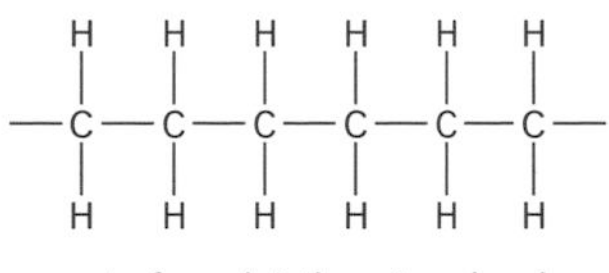

part of a poly(ethene) molecule

Draw a (ring) around the correct answer in each box to complete each sentence.

i Each hydrogen atom is joined to [1 / 2 / 4] other atom(s) .

(1 mark)

ii The bonds between the atoms in the molecule are [covalent. / metallic. / ionic.]

(1 mark)

iii A piece of LDPE consists of many individual polymer chains.

This means it is a [thermosetting / thermosoftening / cross-linked] polymer.

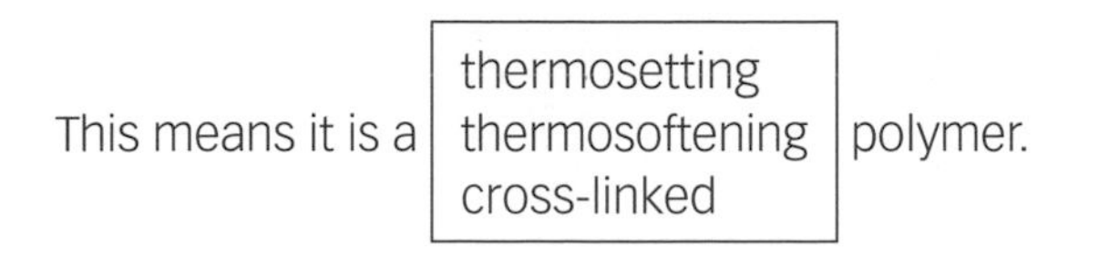

(1 mark)

(Total marks: 5)

2 Read this article, then answer the questions below it.

> Carbon nanotubes are a form of carbon. The carbon atoms are joined together in tiny tubes. The properties of carbon nanotubes are very different to the properties of other forms of carbon, such as graphite.
>
> Carbon nanotubes are very strong when subjected to pulling forces. They are also extremely stiff. They conduct heat well.
>
> A group of scientists have done experiments and found out that carbon nanotubes can enter human cells kept in test tubes. This makes the cells die. Studies on mice and rats suggest that inhaling carbon nanotubes over weeks or months may cause lung problems.

a Draw a (ring) around the correct answer below.

The diameter of a typical nanotube is

1–10 nm **1–10 mm** **1–10 cm** **1–10 m**

(1 mark)

b Carbon nanotubes can be used to reinforce the materials used to make wind turbines.

Give **two** properties of carbon nanotubes that make them suitable for this purpose.

1 ....

2 ....

(2 marks)

c i Identify **two** possible health risks to humans who make or use carbon nanotubes.

1 ....

2 ....

(2 marks)

ii Suggest why we cannot be certain that each of these health problems will occur when a person is exposed to carbon nanoparticles.

....

....

(1 mark)

(Total marks: 6)

3 Rhodium is a metal.

a Small amounts of rhodium are added to platinum metal.

The resulting alloy is very hard.

Explain why the platinum-rhodium alloy is harder than pure platinum.

Your answer should include details of the following:

- How the atoms are arranged in pure platinum and in the platinum–rhodium alloy.
- Why pure platinum is relatively soft.
- Why the platinum–rhodium alloy is harder.

....

....

....

....

(4 marks)

H

b Rhodium is used to coat electrical contacts.

Explain why rhodium is a good conductor of electricity.

....

....

(2 marks)

(Total marks: 6)

6: Atomic structure

Revision objectives

- work out the mass number and atomic number of an atom
- explain what an isotope is
- calculate relative formula mass

Student book references

2.12 Atomic structure
2.13 Masses and moles

Specification key

- C2.3.1

Inside atoms

Atoms are made up of tiny **sub-atomic particles**. In the centre of an atom is its **nucleus**. This is made up of **protons** and **neutrons**. The nucleus is surrounded by **electrons**.

Name of particle	Relative mass	Relative charge
proton	1	+1
neutron	1	0
electron	very small	–1

Representing atoms

You can represent a sodium atom like this:

$$^{23}_{11}Na$$

- The **atomic number** is the number of protons in an atom. The atomic number of sodium is 11.
- The **mass number** is the total number of protons and neutrons in an atom. The atom of sodium above has 11 protons and 12 neutrons. Its mass number is (11 + 12) = 23.

Isotopes

Atoms of the same element can have different numbers of neutrons. This gives them different mass numbers. Atoms of an element that have different numbers of neutrons are called **isotopes**.

Oxygen has three naturally occurring isotopes. The table below shows the number of neutrons in one atom of each isotope.

Isotope	Number of protons	Number of neutrons	Mass number
^{16}O	8	8	16
^{17}O	8	9	17
^{18}O	8	10	18

H Relative atomic mass

The **relative atomic mass, A_r**, of an element compares the mass of atoms of the element with the mass of atoms of the ^{12}C isotope.

Relative atomic mass is an average value for the isotopes of an element. For example, copper has two isotopes:

- About 25% of copper atoms are of the ^{65}Cu isotope.
- About 75% of copper atoms are of the ^{63}Cu isotope.

The relative atomic mass of copper is 63.5. This is an average of the masses of the two isotopes, taking into account their relative amounts.

Relative formula mass

The formula of a substance tells you the number of each type of atom in the substance. The chemical dopamine is released by the brain when people fall in love. Its formula is $C_8H_{11}NO_2$. The formula tells you that one molecule of dopamine is made up of:

- 8 carbon atoms
- 11 hydrogen atoms
- 1 nitrogen atom
- 2 oxygen atoms.

The **relative formula mass (M_r)** of a substance is the sum of the relative atomic masses of the atoms in the numbers shown in the formula. You can work it out by adding together all the A_r values for the atoms in the formula. The relative formula mass of dopamine is

$$(12 \times 8) + (1 \times 11) + 14 + (16 \times 2) = 153.$$

Moles

The relative formula mass of a substance compares the mass of the substance to the mass of an atom of ^{12}C. You can also compare the masses in grams. The relative formula mass of a substance, in grams, is known as one **mole** of the substance. So the mass of one mole of carbon atoms is 12 g. The mass of one mole of dopamine molecules is 153 g.

Key words

sub atomic particle, nucleus, proton, neutron, electron, atomic number, mass number, isotope, relative atomic mass A_r, relative formula mass M_r, mole

Exam tip

AQA

Practise working out mass numbers and atomic numbers, and calculating relative formula masses.

Questions

1. Give the atomic numbers and mass numbers of these atoms:
   $^{14}_{7}N$ $^{56}_{26}Fe$ $^{80}_{35}Br$
2. Explain the meaning of the word 'isotope'.
3. Calculate the relative formula mass of lead nitrate, $Pb(NO_3)_2$. Use data from the periodic table to help you.

Revision objectives

- calculate the percentage of an element in a compound
- H calculate empirical formulae, masses from equations, and percentage yields

Student book references

2.16 Using equations
2.17 Calculating yield

Specification key

- C2.3.3

Percentage by mass

To calculate the **percentage by mass** of an element in a compound, use the equation:

$$\text{percentage by mass} = \frac{\text{relative mass of element in compound}}{\text{relative formula mass}} \times 100$$

Worked example

Q What is the percentage by mass of nitrogen in ammonium nitrate (NH_4OH)?

A The relative formula mass, $M_r = 14 + (1 \times 4) + 16 + 1 = 35$
The relative mass of nitrogen in the formula = 14
Percentage of nitrogen in ammonium nitrate $= \frac{14}{35} \times 100 = 40\%$

H Empirical formulae

The **empirical formula** of a compound gives the relative number of the atoms of each element that are in the compound.

Worked example

Q A sample of a compound contains 0.20 g of hydrogen and 1.6 g of oxygen. What is its formula?

A

	hydrogen	oxygen
mass of each element, in g	0.20	1.6
A_r from periodic table	1	16
mass divided by A_r	0.2	0.1
simplest ratio	2	1

The formula of the compound is H_2O.

H Calculating masses of reactants and products from equations

In a chemical reaction, the total mass of reactants is equal to the total mass of products. This means you can calculate the masses of reactants and products from balanced symbol equations.

Worked example

Q 32 g of methane burns in air. What are the maximum masses of carbon dioxide and water that can be produced?

A $CH_4 + 2O_2 \rightarrow CO_2 + 2H_2O$

M_r of $CH_4 = 12 + (1 \times 4) = 16$

M_r of $CO_2 = 12 + (16 \times 2) = 44$

M_r of $H_2O = (1 \times 2) + 16 = 18$

Use ratios to work out the answer:

16 g of methane makes 44 g of carbon dioxide.

So 32 g of methane makes (44 × 2) = 88 g of carbon dioxide and (18 × 2) = 36 g of water.

Key words

percentage by mass, empirical formula, reversible, actual yield, maximum theoretical yield, percentage yield

Yield

In a chemical reaction, atoms cannot be gained or lost. But you don't always make the calculated amount of product. This may be because:

- you lose some of the product when separating it from the reaction mixture – in filtering, for example
- a reactant reacts in an unexpected way, for example, when burning in air, a substance might react with nitrogen as well as oxygen
- the reaction may be **reversible** – the products react to make the original reactants. For example:
  ammonium chloride ⇌ ammonia + hydrogen chloride

The amount of product formed in a reaction is the **actual yield**. You can compare the actual yield to the **maximum theoretical yield**, calculated from the balanced symbol equation, to find the **percentage yield** of a reaction.

Worked example

Q Jasmine heated a piece of magnesium in air. She calculated a maximum theoretical yield for magnesium oxide of 8 g. But the actual yield was only 6 g. Calculate the percentage yield.

A $\text{percentage yield} = \dfrac{\text{actual yield}}{\text{maximum theoretical yield}} \times 100$

$= \dfrac{6\,g}{8\,g} \times 100 = 75\%$

Exam tip AQA

Practise finding relative atomic masses from different elements in the periodic table.

Questions

1. Calculate the percentage of oxygen in magnesium oxide, MgO.
2. Explain why the maximum theoretical yield of a product is not always obtained in a chemical reaction.
3. **H** A sample of a compound contains 1.20 g of carbon and 0.4 g of hydrogen. What is its formula?

Revision objectives

- describe how to use paper chromatography and instrumental techniques to separate mixtures and identify substances

Student book references

2.14 Chemical detectives

2.15 Identifying chemicals

Specification key

- **C2.3.2**

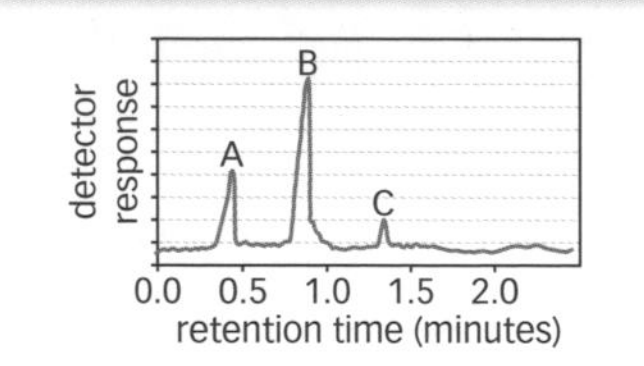

Key words

retention time, mass spectrometer, molecular ion peak

H A mass spectrometer produces a mass spectrograph like this for each compound that it analyses.

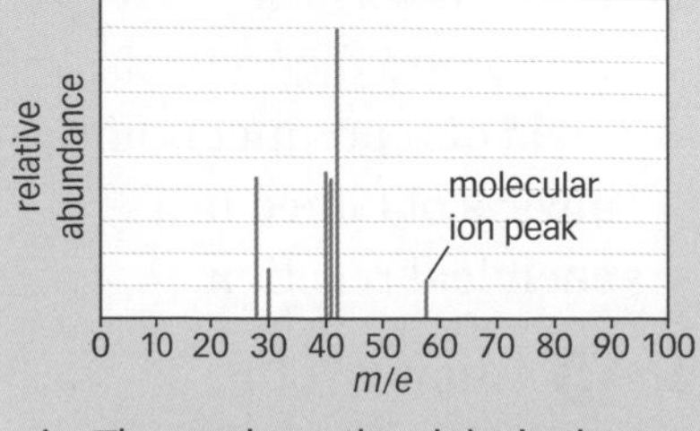

▲ The peak on the right is the molecular ion peak. The mass of the molecular ion gives the relative molecular mass of the compound.

Chemical analysis

You can use paper chromatography to separate artificial food colours.

The chromatogram shows that the brown colouring is a mixture of blue, red, and yellow dyes. ▶

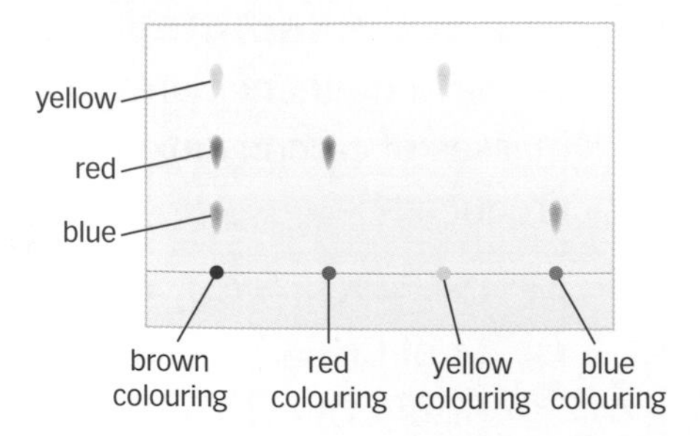

Gas chromatography mass spectroscopy (GC-MS)

GC-MS is an instrumental method of analysis. Instrumental methods identify small samples. They are sensitive, accurate, and quick.

Gas chromatography separates mixtures. The diagram shows how it works.

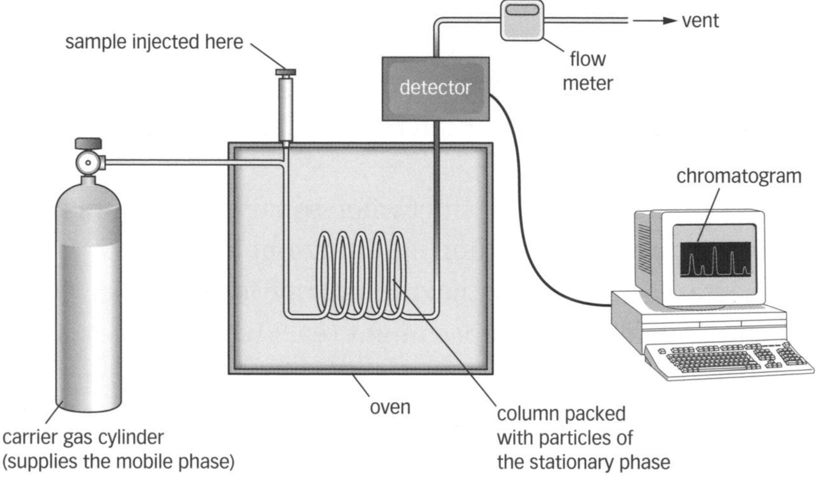

▲ Inside a gas chromatography instrument.

The gas chromatogram on the left (above) shows that:

- there are three peaks, so the mixture has three compounds
- compound A travelled most quickly, and left the column first. This means that compound A has the shortest **retention time**. Retention times help identify substances.
- peak B has the greatest area, so compound B is present in the mixture in the greatest amount.

The gas chromatography apparatus may be connected to a **mass spectrometer**. This identifies the separated compounds.

Questions

1 List **three** advantages of instrumental analysis methods.

2 Explain how gas chromatography apparatus separates the compounds of a mixture.

Working to Grade E

1 Draw lines to match each type of particle to its relative mass.

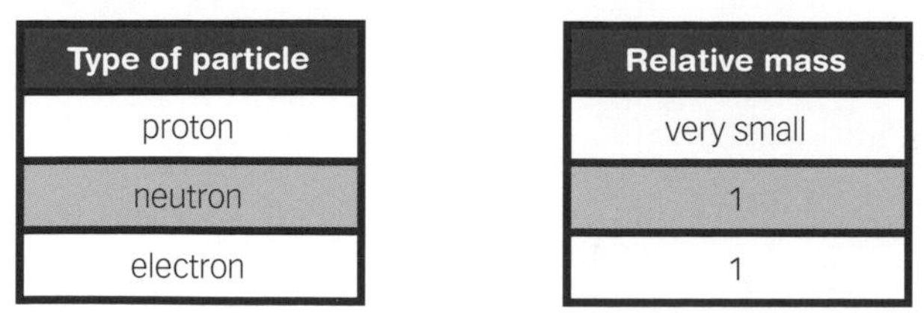

Type of particle
proton
neutron
electron

Relative mass
very small
1
1

2 Tick the boxes to show the benefits of instrumental methods of chemical analysis.

- **a** They are quick. ☐
- **b** The instruments are expensive. ☐
- **c** They are accurate. ☐
- **d** They can identify tiny amounts of substances. ☐
- **e** People need to be highly trained to operate the instruments. ☐

3 The statements below describe the steps in using GC-MS to separate and identify compounds. Write the letters of the steps in the best order.

- **A** The different vapours in the mixture travel through the column, which is packed with a solid material, at different speeds.
- **B** A liquid mixture of unknown compounds is injected into the gas chromatography instrument.
- **C** The vapours leave the column at different times.
- **D** A carrier gas mixes with the vapours.
- **E** This separates the vapours in the mixture.
- **F** The mass spectrometer identifies the substances as they leave the chromatography column.
- **G** The liquid mixture is heated so that it becomes a mixture of vapours.

4 Highlight the statement or statements below that are true. Then write corrected versions of any that are false.

- **a** Isotopes are atoms of the same element that have different numbers of neutrons.
- **b** The relative formula mass of a substance, in grams, is called one squirrel of that substance.
- **c** The theoretical yield of a substance in a chemical reaction is always the same as, or greater than, the actual yield.

Working to Grade C

5 Give the atomic numbers and mass numbers of the atoms listed below.

a $^{40}_{18}Ar$ **b** $^{55}_{25}Mn$ **c** $^{65}_{30}Zn$

6 Calculate the masses of one mole of:

- **a** paracetamol, $C_8H_9NO_2$
- **b** aspirin, $C_9H_8O_4$

7 Calculate the percentage by mass of:

- **a** potassium in potassium cyanide, KCN
- **b** lithium in lithium carbonate, Li_2CO_3
- **c** nitrogen in serotonin, $C_{10}H_{12}N_2O$

8 Use the gas chromatograph to answer the questions.

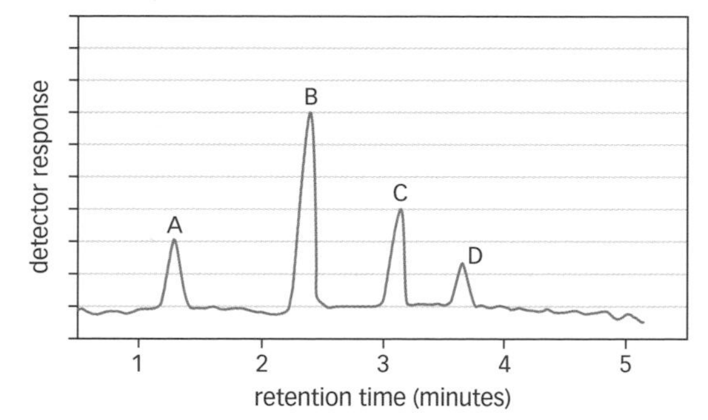

- **a** How many compounds were in the mixture?
- **b** Which substance has the longest retention time?
- **c** Which substance travelled most quickly?

9 Give **three** reasons to explain why it is not always possible to obtain the maximum theoretical yield of a product in a chemical reaction.

Working to Grade A*

10 The diagram below is a simplified mass spectrum of a substance. What is the molecular mass of the substance?

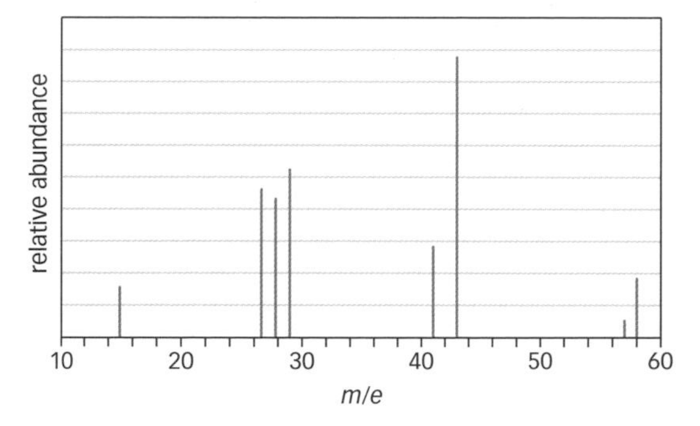

11 Write a definition for the relative atomic mass of an element, A_r.

12 Calculate the formulae of samples of compounds that contain:

- **a** 3.2 g of sulfur and 3.2 g of oxygen
- **b** 2.4 g of carbon and 0.4 g of hydrogen
- **c** 2.3 g of sodium, 1.4 g of nitrogen and 4.8 g of oxygen.

13 Calculate the maximum theoretical yield of carbon dioxide, if a student heats 10.0 g of calcium carbonate. The equation for the reaction is $CaCO_3 \rightarrow CaO + CO_2$

14 Miss Corner heats a known mass of potassium in chlorine. The maximum theoretical yield of potassium chloride is 1.5 g. The actual yield is 1.0 g. Calculate the percentage yield.

Examination questions
Atomic structure, analysis, and quantitative chemistry

1 a An atom of the element zirconium may be represented like this:

$$^{91}_{40}Zr$$

i Give the mass number of this zirconium atom.

....

(1 mark)

ii Give the number of protons in an atom of zirconium.

....

(1 mark)

b Two more atoms of zirconium are represented like this:

$$^{92}_{40}Zr \text{ and } ^{94}_{40}Zr$$

i Give the number of neutrons in each of these zirconium atoms.

$^{92}_{40}Zr$ ....

$^{94}_{40}Zr$ ....

(2 marks)

ii Draw a (ring) around the correct answer to complete the sentence below.

Different atoms of zirconium have different numbers of neutrons.

These atoms are called [molecules / ions / isotopes] of zirconium.

(1 mark)

(Total marks: 5)

2 Morphine is made in the body after the injection of the drug heroin.

a The formula of morphine is $C_{17}H_{19}NO_3$.

Calculate its relative formula mass, M_r.

....

....

(2 marks)

b The formula of heroin is $C_{21}H_{23}NO_5$.

Its relative formula mass, M_r, is 369.

i What is the mass of one mole of heroin?

Include the unit in your answer.

....

(1 mark)

ii Calculate the percentage by mass of oxygen in heroin.

....

....

....

(2 marks)

c Morphine can be detected in the hair of heroin users.

The diagram shows part of a simplified gas chromatogram obtained by analysing the hair of a heroin user.

The heroin user had also taken the drug codeine.

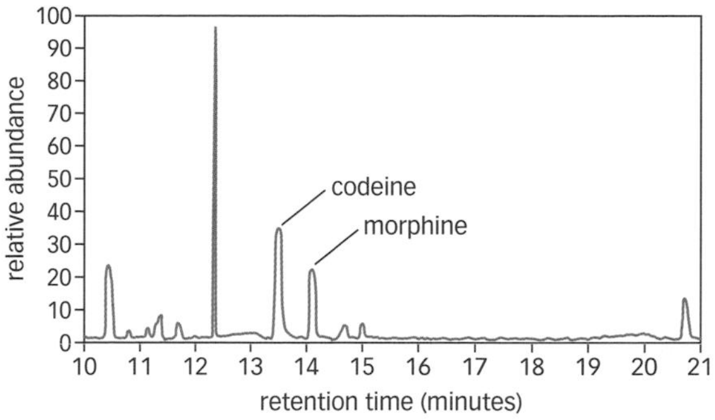

i Which travels more slowly through the column in the gas chromatography apparatus, morphine or codeine?

Give a reason for your decision.

....

....

(1 mark)

ii Draw a (ring) around the correct answer to describe what happens in the column of the gas chromatography apparatus.

The mixture is separated.

The relative molecular masses of the compounds in the mixture are measured.

Liquids are vapourised.

(1 mark)

(Total marks: 7)

H **3** A chemist heated 1.2 g of magnesium in air.

She used the equation below, and data from the periodic table, to calculate the maximum theoretical yield of magnesium oxide as 2.0 g.

$$2Mg(s) + O_2(g) \rightarrow 2MgO(s)$$

She actually obtained only 1.5 g of magnesium oxide.

a Calculate the percentage yield of magnesium oxide.

....

....

(2 marks)

b Suggest why the actual yield of magnesium oxide was less than the maximum theoretical yield.

....

....

(1 mark)

(Total marks: 3)

How Science Works
Structures, properties, and uses

Societal aspects of scientific evidence

Decisions about scientific issues are not usually based on evidence alone. Other factors, such as those relating to ethical, social, economic or environmental concerns, are also taken into account when evaluating the impacts of new developments.

Materials for devices to keep blood vessels open

Skill – Analysing the facts and making deductions

In this question you will be assessed on using good English, organising information clearly, and using specialist terms where appropriate.

Read the information below and answer the questions that follow.

A doctor may insert a stent into a blood vessel that has become narrow or blocked. The stent helps to keep the blood vessel open.

Some stents are made from stainless steel. Others are made from a shape-memory alloy called Nitinol. At body temperature, Nitinol stents change shape to match the shape of the blood vessels they are holding open. Stainless steel does not change shape in this way. Nitinol stents return to their original shape after being squashed. Stainless steel stents do not.

Research suggests that blood clots are less likely to form on Nitinol stents than on stainless steel ones.

Nitinol is an alloy of nickel and titanium. Some people are allergic to nickel, and it is possible that, in these people, new blockages might form in the blood vessel near the stent. If nickel compounds get into the blood stream, the risk of cancer may increase.

However, if the surface of a Nitinol stent is treated properly, a layer of titanium dioxide forms on its surface. This means that nickel is very unlikely to get into the bloodstream of a person with a stent.

Stainless steel is mainly iron. It contains small amounts of other metals. Scientists have shown that if people are allergic to these metals, new blockages might form in the blood vessel near the stent.

1 Use the information in the box to evaluate the use of stainless steel compared with Nitinol as a material for making stents.

This question includes the word *evaluate.* To answer it, you will need to write down some advantages and disadvantages of making stents from Nitinol, and some advantages and disadvantages of making stents from stainless steel. You will then need to compare the advantages and disadvantages, and write down which material is better for stents, and why.

Before you start writing your answer, you may find it helpful to organise your ideas in a table like the one below.

	Advantages	Disadvantages
Nitinol		
Stainless steel		

Once you've organised your ideas like this, you can decide which you think is the better material, and why.

Of course, there is no one 'correct answer' to this question. You can get full marks whatever your decision, provided you have stated the advantages and disadvantages clearly, and given reasons linked to these for your final decision.

2 Suggest one economic factor that a hospital might consider before deciding which of the two types of stent to offer its patients.

To answer this question, you will need to think of a factor that is related to money.

Skill – Evaluating bias

3 A student makes notes about four scientists who have done research about stents and the materials they are made from.

Anita Smith – 22 years old, funding herself through university.
Professor Nadeem Hanif – university scientist with 25 years experience. Government money funds his research.
Dr Bernard Anning – scientist with 30 years experience. Works for company that makes Nitinol.
Dr Rachel Hooper – hospital doctor who has inserted many stainless steel and Nitinol stents into patients over the past two years, and has collected data on their health after having the stents. Research funded by company that makes stainless steel stents.

a The student says that she has greater trust in the research findings of Anita Smith than of Rachel Hooper. Suggest **one** reason for this opinion.

b Using only the information in the box above, name the scientist whose evidence you think other scientists should pay most attention to. Give a reason for your decision.

In answering parts a and b you will need to consider whether the evidence from any of the scientists might be biased, and if so, why. You also need to think about the status of the researchers – for example, who is more experienced? For part b there is no one correct answer – the examiner is more interested in the reasons you give to back up your decision.

AQA Upgrade

Answering a question using formulae

The compound strontium chloride is used in red fireworks. It is also added to seawater aquaria, where it is used by tiny sea creatures to make their skeletons.

1 This diagram shows the electronic structure of a chlorine atom. Draw the structure of the chloride ion, Cl^-. *(2 marks)*

2 The formula of the strontium ion is Sr^{2+}. Give the formula of strontium chloride. *(1 mark)*

3 Explain why solid strontium chloride does not conduct electricity, but a solution of strontium chloride does conduct electricity. *(4 marks)*

QUESTION

B–A*

1

2 $SrCl_2$

3 In solid strontium chloride, there are strong electrostatic forces in all directions between the oppositely charged ions. The ions are not free to move to carry an electric current. In a solution, the charged particles are free to move and carry the current. So a solution of strontium chloride does conduct electricity.

Examiner: This answer gains seven marks out of seven. All parts are answered correctly.

The explanation given in answer to question 3 is accurate and detailed.

D–C

1

2 Sr_2Cl

3 The solid doesn't conduct because it has no ions that are free to move. The solution has moving ions so it conducts.

Examiner: This answer gains four marks out of seven.

Question 1 is answered correctly, and gains two marks.

Question 2 is incorrect, and gains zero marks. The candidate has not checked that the total number of positive charges on the ions shown in the formula is equal to the total number of negative charges on the ions shown in the formula.

The student gains two marks out of four for question 3, since each part of it is answered correctly, but in insufficient detail.

G–E

1

2 $SrCl$

3 The solid does not conduct because it is all held together by strong covalent bonds, but the solution does because the ions are free to move.

Examiner: This answer gains just one mark out of seven.

The structure given for question 1 is incorrect, since the highest occupied energy level contains nine electrons, instead of eight.

The formula for question 2 is also incorrect – two Cl^- ions are needed to balance the two positive charges on the Sr^{2+} ion.

One mark is awarded for the second half of the sentence that answers question 3. The first part of the answer is incorrect – solid strontium chloride is held together by ionic bonds, not covalent ones.

Part 1 Course catch-up
Structures, properties, and uses

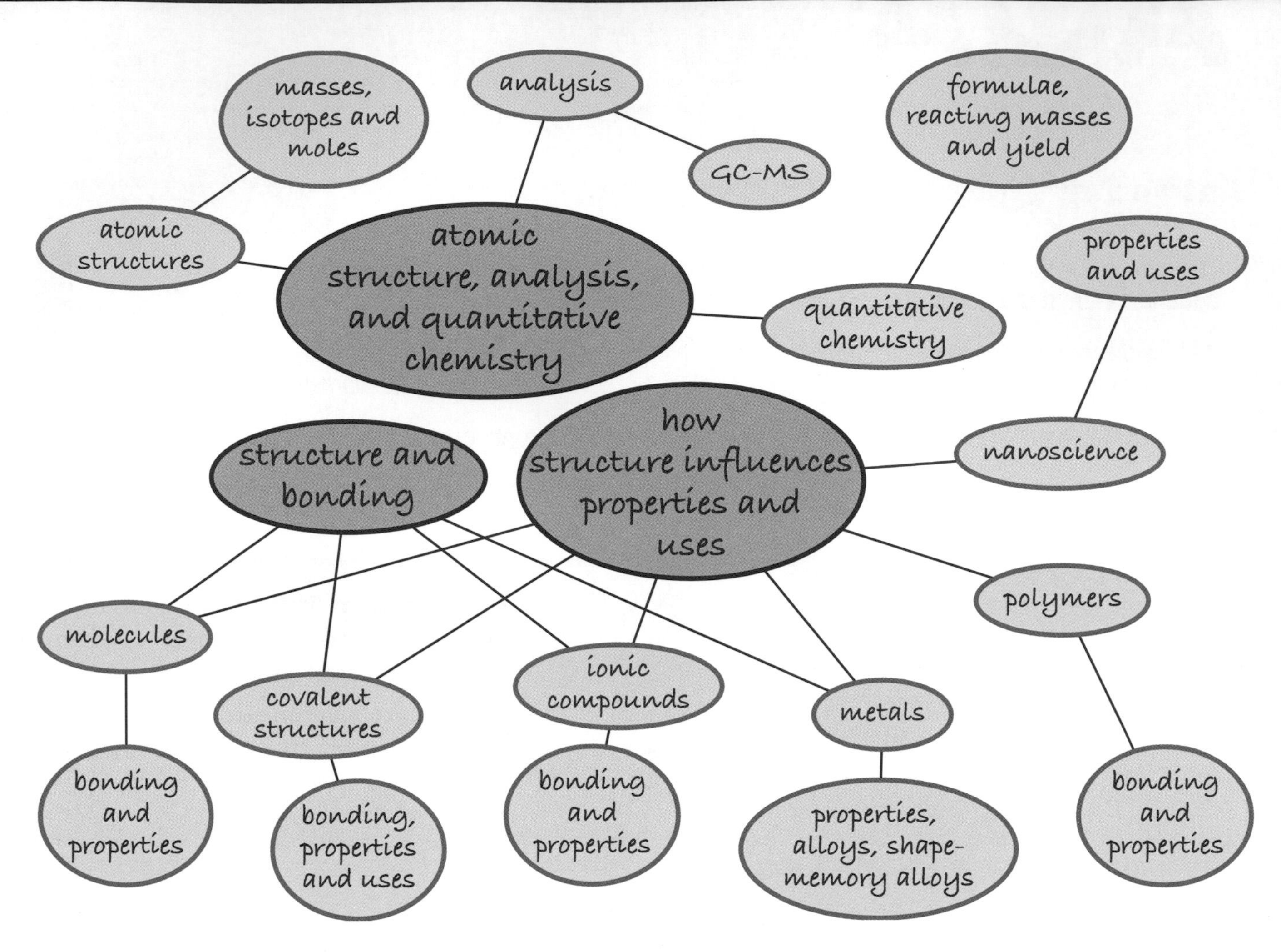

Revision checklist

- Chemical bonding involves transferring or sharing electrons in the highest occupied energy levels of atoms to achieve the stable electronic structure of a noble gas.
- Ions are formed when electrons transfer from one atom to another.
- Ionic compounds are held together by strong electrostatic forces of attraction. This gives them high melting and boiling points. They conduct electricity when molten or in solution.
- Strong covalent bonds form when atoms share electrons. Some covalently bonded substances consist of simple molecules. Others have giant structures of atoms.
- Substances with giant covalent structures include forms of carbon, such as graphite and diamond, and silicon dioxide. They have high melting points.
- Metals have giant structures of atoms arranged in layers in a regular pattern. The layers slide over each other, so metals are bendy.
- Metals have delocalised electrons in their structures, so they can conduct heat and electricity.
- Most alloys are made of two or more metals. Different sized atoms distort the layers, so alloys are harder than pure metals.
- Shape-memory alloys such as Nitinol return to their original shape after being deformed.
- Nanoparticles are structures of between 1 and 100 nm in size. They have different properties to the same material in bulk. They may be used in new computers, catalysts, and cosmetics.
- The total number of protons and neutrons in an atom is its mass number.
- Atoms of the same element with different numbers of neutrons are isotopes.
- The relative atomic mass of an element compares the mass of atoms of the element with the ^{12}C isotope.
- The relative formula mass (M_r) of a compound is the sum of the relative atomic masses of the atoms in the numbers shown in the formula.
- Gas chromatography linked to mass spectroscopy (GC-MS) is an instrumental method of analysis. It is accurate, sensitive, and quick.
- Information from formulae can be used to calculate the percentage of an element in a compound.
- Reacting masses can be calculated from chemical equations.
- In a chemical reaction, the actual yield of a product may be less than the maximum theoretical yield. The percentage yield can be calculated from these values.

Reaction rates

The **rate of reaction** is a measure of how quickly a reaction happens. To work out the rate of a reaction, you need to do experiments to find out how quickly products are made or reactants are used up.

Following reactions

This equation shows how magnesium reacts with hydrochloric acid.

magnesium + hydrochloric acid → magnesium chloride + hydrogen

$$Mg + 2HCl \rightarrow MgCl_2 + H_2$$

You can follow the reaction by measuring the volume of hydrogen made as the reaction happens.

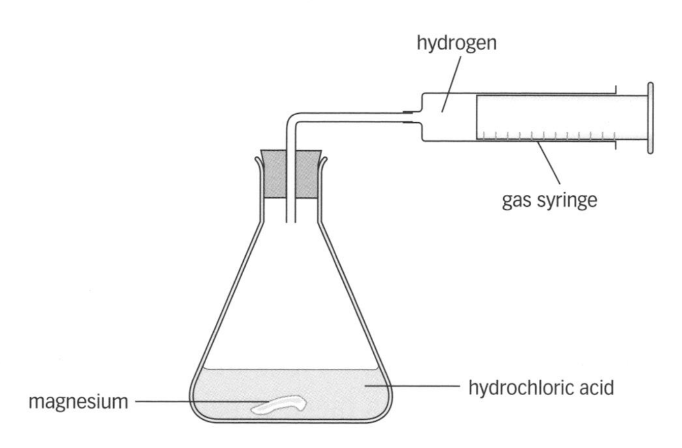

The graph on the right shows the volume of hydrogen made during the reaction.

- At first, the gradient is steep, showing that the reaction is fast.
- The gradient gets less steep over time, showing that the reaction slows down.

Calculating rates

You can use the equation below to calculate the rate of the reaction of magnesium with hydrochloric acid.

$$\text{rate of reaction} = \frac{\text{amount of product made}}{\text{time}}$$

From the graph, the rate of the reaction for the first minute

$$= \frac{25\,\text{cm}^3}{1\,\text{min}} = 25\,\text{cm}^3/\text{min}$$

Over the first three minutes, the average reaction rate

$$= \frac{40\,\text{cm}^3}{3\,\text{min}} = 13.3\,\text{cm}^3/\text{min}$$

Revision objectives

- explain the meaning of the term 'rate of reaction'
- use data, equations, and graphs to calculate reaction rates
- describe and explain the effect of temperature on reaction rate

Student book references

2.18 How fast?
2.19 Speeding up reactions – temperature

Specification key

✔ C2.4.1 a – c

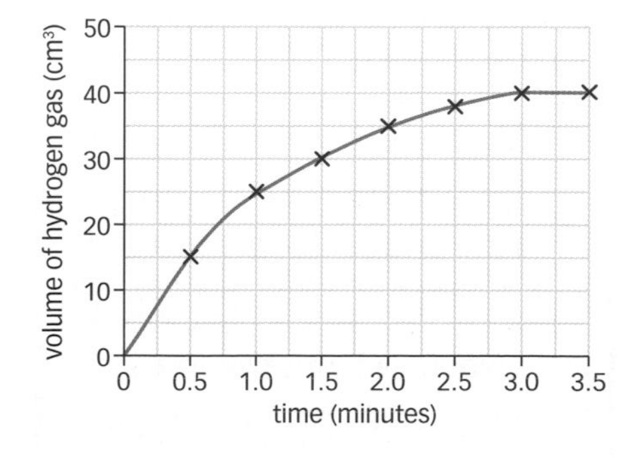

Key words

rate of reaction, collide, activation energy, successful collision

If you have data for the amount of reactant, you can use the equation below to calculate reaction rates:

$$\text{rate of reaction} = \frac{\text{amount of reactant used}}{\text{time}}$$

Collisions and activation energy

Reactions can only happen when reactant particles **collide**, or hit each other. The colliding particles must have enough energy to react.

The minimum amount of energy that particles need in order to react is the **activation energy**.

Temperature and reaction rate

Changing the reaction conditions changes the rate of a reaction.

Increasing the temperature increases the rate of a reaction. This is because higher temperatures make particles in a reacting mixture move faster. This makes their collisions:

- more frequent
- more energetic.

The more energetic a collision, the greater the likelihood of it leading to a reaction, and so being a **successful collision**.

Sodium thiosulfate reacts with hydrochloric acid according to the equation below. The sulfur formed in the reaction is a solid. It obscures a cross drawn on a piece of paper placed beneath the reaction flask.

sodium thiosulfate + hydrochloric acid → sodium chloride + water + sulfur dioxide + sulfur

The graph below shows the link between temperature and reaction time for the reaction. The shorter the reaction time, the faster the rate of reaction.

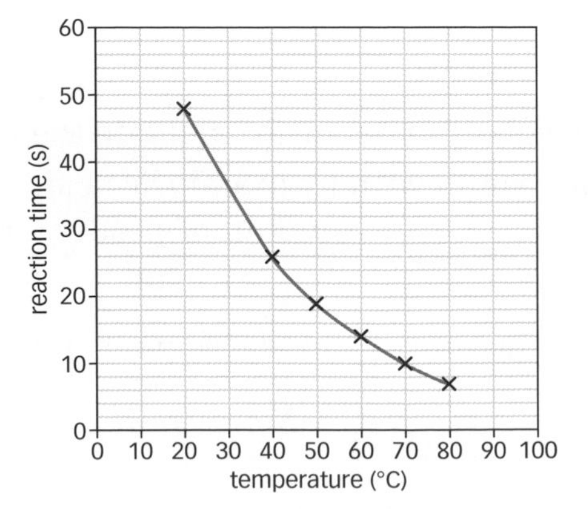

Exam tip

AQA

Practise using the rate of reaction equations, and interpreting graphs.

Questions

1. Describe the link between temperature and reaction rate shown by the graph to the right.
2. Use data from the graph on page 91 to calculate the average rate of reaction over the first two minutes of the reaction.
3. Explain why the volume does not increase after three minutes for the graph on page 91.

Pressure

You can speed up reactions involving gases by increasing the pressure.

Increasing the pressure of a mixture of gases makes the particles more crowded. This means they collide more frequently. Increasing the frequency of collisions increases the rate of the reaction.

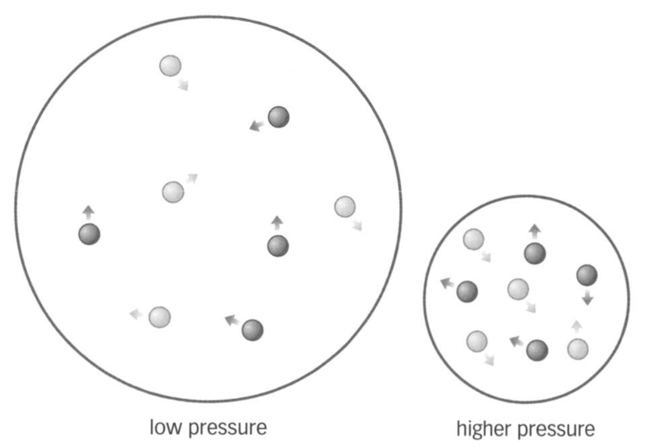

▲ Gas particles collide more frequently at higher pressures.

Revision objectives

- ✓ describe and explain the effects of changing concentration, pressure and surface area on rates of reaction
- ✓ explain what catalysts do and why they are important in industry

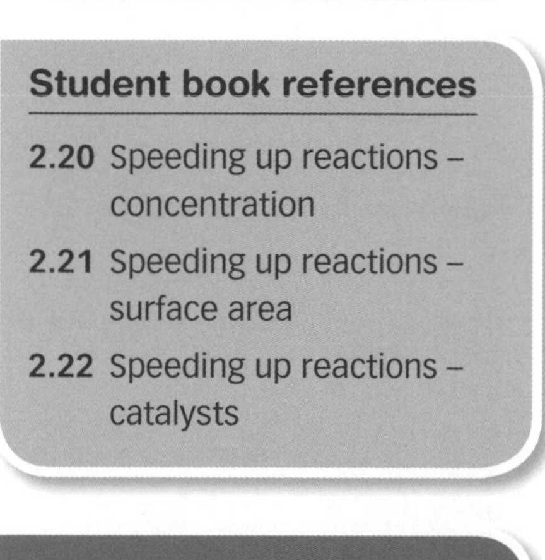

Student book references

2.20 Speeding up reactions – concentration

2.21 Speeding up reactions – surface area

2.22 Speeding up reactions – catalysts

Specification key

✓ **C2.4.1 d – h**

Concentration

You can speed up a reaction involving a solution by increasing its concentration.

The more concentrated a solution, the greater the number of solute particles dissolved in a certain volume of the solution, and the more crowded the particles. Increased crowding leads to more frequent collisions, and a faster reaction.

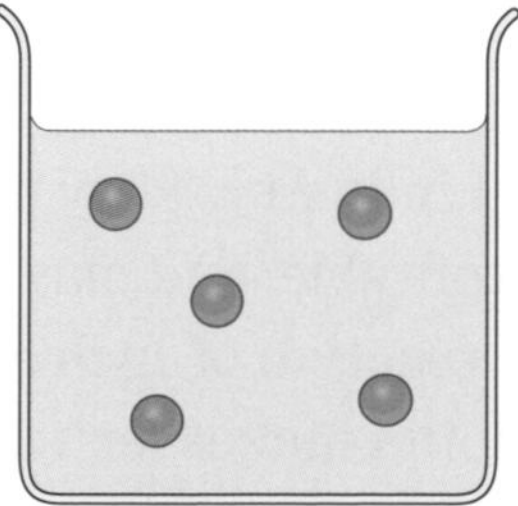

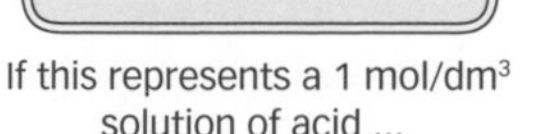
If this represents a 1 mol/dm^3 solution of acid ...

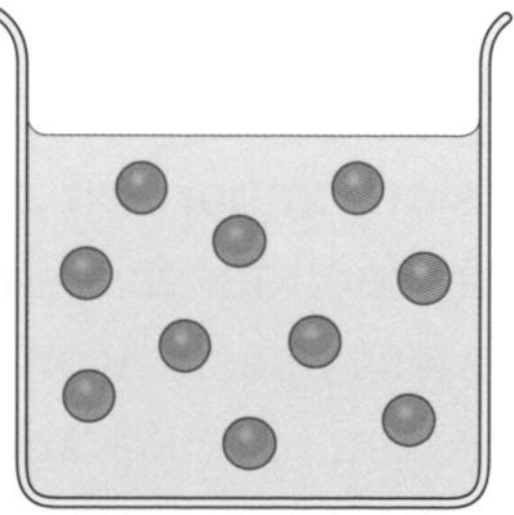

then this represents a 2 mol/dm^3 solution of the same acid.

▲ There are double the number of acid particles in the same volume of water. Water particles are not shown on the diagrams.

The graph opposite shows the link between reaction time and acid concentration for the reaction of hydrochloric acid with magnesium ribbon. Remember – the shorter the reaction time, the faster the reaction.

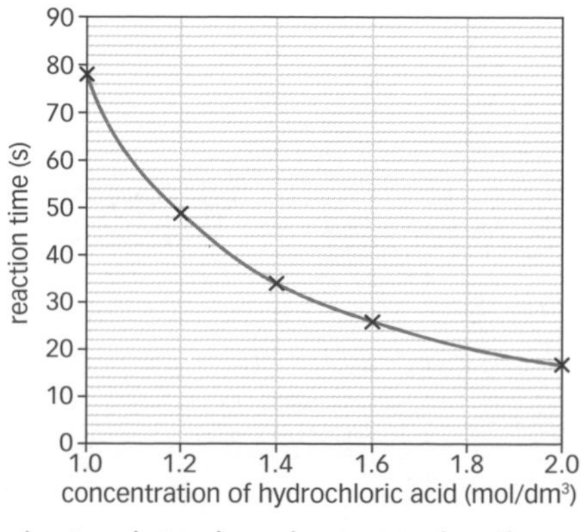

▲ Graph to show how reaction time varies with concentration.

Key words

surface area, catalyst, catalyses

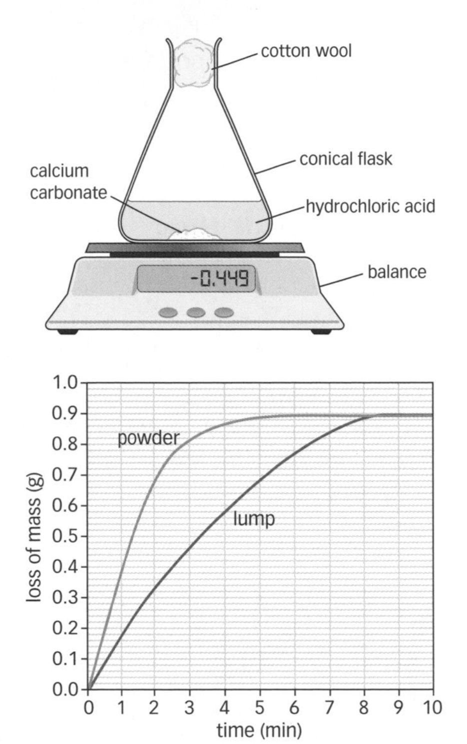

▲ The loss in mass during the reaction of calcium carbonate with hydrochloric acid.

Surface area

A reaction involving a powder happens faster than a reaction involving a lump of the same reactant. A powder has a bigger **surface area** than a lump of the same mass. This is because particles that were inside the lump become exposed on the surface when it is crushed.

The bigger the surface area, the greater the frequency of collisions, and the faster the reaction.

The equation below summarises the reaction of hydrochloric acid with calcium carbonate.

hydrochloric acid + calcium carbonate → calcium chloride + water + carbon dioxide

You can use the apparatus on the left to follow the reaction by measuring the decrease in mass as carbon dioxide gas is made and given off.

The graph shows that the reaction is faster when powdered calcium carbonate is used.

Catalysts

You can increase the rate of many reactions by adding a **catalyst** to the reaction mixture. A catalyst increases the rate of a reaction without being used up in the reaction.

For example, hydrogen peroxide in solution breaks down very slowly to form water and oxygen:

hydrogen peroxide → water + oxygen

$$2H_2O_2 \rightarrow 2H_2O + O_2$$

Adding powdered manganese(IV) oxide **catalyses**, or speeds up, the reaction, making the hydrogen peroxide decompose more quickly.

Catalysts are important in the chemical industry. They make chemical reactions fast enough to be profitable, and may reduce energy costs. Iron catalyses the reaction of hydrogen with nitrogen to make ammonia. Ammonia makes fertilisers and explosives.

Exam tip AQA

Remember, as the reaction rate increases, the reaction time gets less.

Questions

1 List **four** factors that affect reaction rate.

2 Use ideas about collisions to explain why increasing the surface area of a solid reactant increases reaction rate.

3 Explain what catalysts do, and why they are important in industry.

Questions
Rates of reaction

Working to Grade E

1 Tick the boxes to show which of the following changes increase the rate of reaction.
   a Increasing the temperature ☐
   b Diluting any solutions present ☐
   c Increasing the gas pressure ☐
   d Decreasing the surface area of any solids ☐

2 What is the activation energy of a reaction? Tick the best definition in the table below.

Definition	Tick (✓)
The temperature change for the reaction.	
The energy change for the reaction.	
The maximum energy the particles must have to react.	
The minimum energy the particles must have to react.	

3 Write a definition of the word 'catalyst'.

4 A student adds hydrochloric acid to calcium carbonate (marble chips) in the apparatus below.

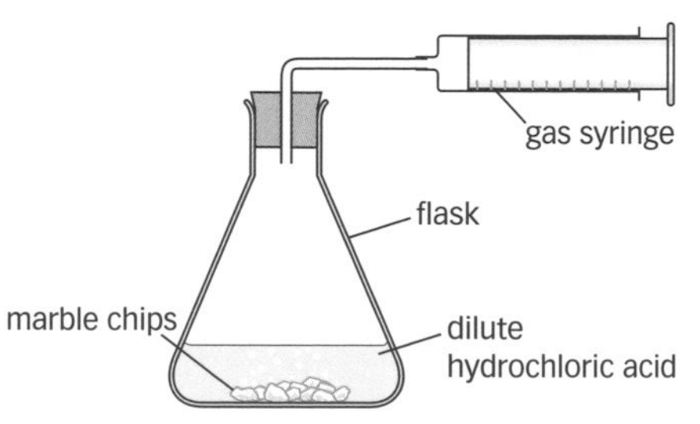

One of the products of the reaction is carbon dioxide gas. Every minute, the student records the volume of carbon dioxide gas in the gas syringe. The graph below shows the results.

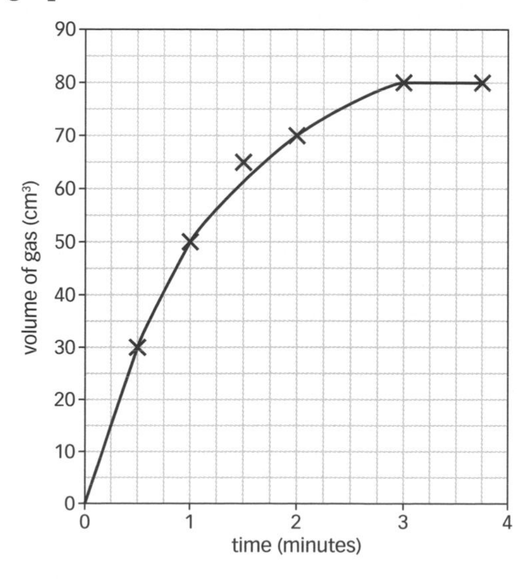

a Which description best describes how the volume of gas changes with time? Tick **one** description.

Description	
The volume of gas increases quickly at first, and then more slowly.	
The volume of gas increases quickly.	
The volume of gas increases slowly at first, and then more quickly.	
The volume of gas increases quickly.	

b Use data from the graph and the equation in the box below to calculate:
   i the rate of reaction in the first minute.
   ii the rate of reaction in the second minute.

$$\text{Rate of reaction} = \frac{\text{volume of product formed}}{\text{time}}$$

c Add a line to the graph above to show how the volume of gas would change with time if the student repeated the investigation at a higher temperature.

Working to Grade C

5 Use ideas about particles to explain why

a increasing the temperature of a reaction increases its rate.

b increasing the concentration of reactants in solution increases the rate of reaction.

6 A chemical company makes ammonia. Suggest why it uses a catalyst to speed up the reaction.

7 A student sets up the apparatus below.

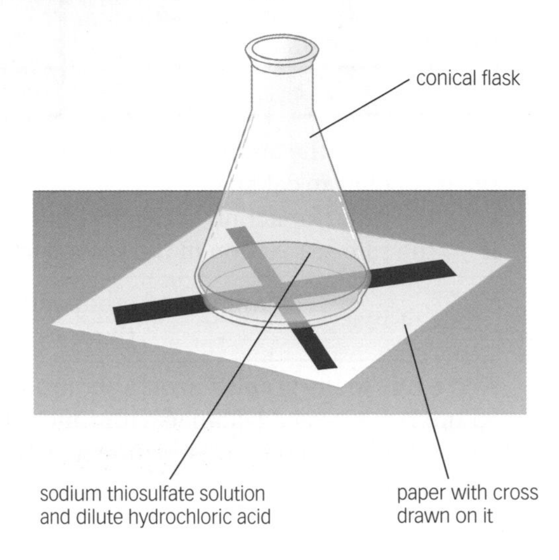

He measures the time taken for the cross below the flask to disappear. He repeats the investigation at a total of five temperatures. This table summarises his results.

Temperature (°C)	Time for cross to disappear (s)
20	400
30	200
40	100
50	47
60	26
70	60

a Identify the independent variable.

b Identify **three** control variables.

c Identify the range of the dependent variable.

d Plot the results on a graph and draw a line of best fit.

e Identify the anomalous result.

f Describe the pattern shown on the graph.

8 A student monitors the reaction of big lumps of calcium carbonate with dilute hydrochloric acid. Every 30 seconds, she measures the total volume of gas that has been formed.
She repeats the procedure with smaller pieces of calcium carbonate, and then with calcium carbonate powder. She keeps all the other reaction conditions the same.
This graph summarises the student's data.

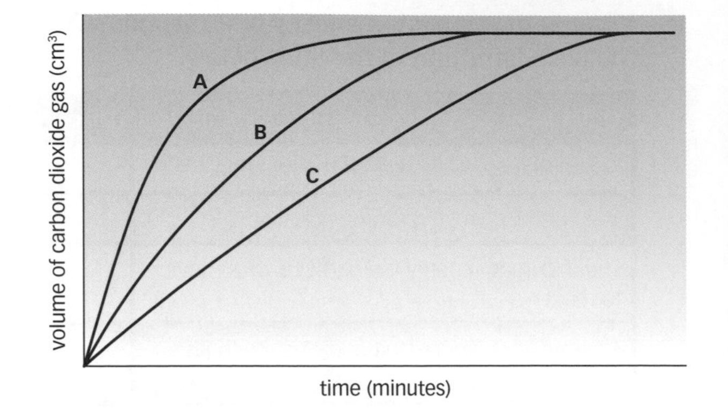

a Which curve on the graph represents the reaction with the biggest pieces of calcium carbonate?

b Explain why the total volume of carbon dioxide made is the same each time.

Examination questions
Rates of reaction

1 A student tested the hypothesis that there is a link between temperature and the rate of a reaction.

He added magnesium ribbon to dilute hydrochloric acid.

He measured the time for the magnesium ribbon to completely react, and disappear.

He repeated the investigation at a total of five different temperatures.

conical flask
magnesium ribbon
hydrochloric acid

a List **two** variables that must be controlled in order to obtain valid results.

1 ....

2 ....

(2 marks)

b The student plotted his results on the graph below. Use the graph to answer the questions that follow.

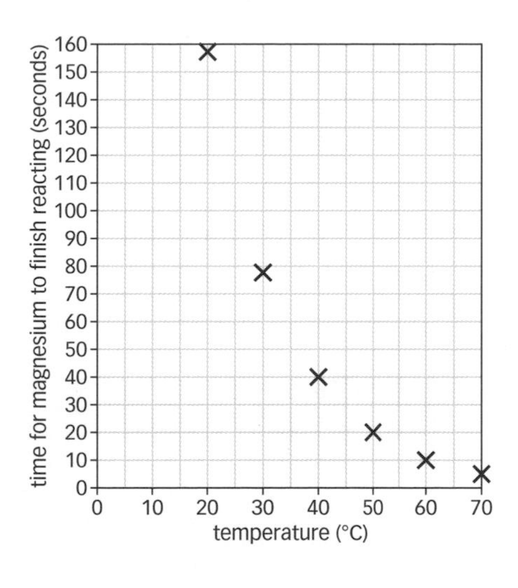

i Draw a line of best fit on the graph. *(1 mark)*

ii Use the graph to predict the time for the magnesium to disappear at 45 °C.

....

(1 mark)

iii Describe the pattern shown by the graph.

....

....

(2 marks)

iv Use ideas about particles to explain the link between temperature and reaction rate.

....

....

(2 marks)

(Total marks: 8)

11: Exothermic and endothermic reactions

Revision objectives

- explain what exothermic and endothermic reactions are, and give examples of each
- recall that reactions that are endothermic in one direction are exothermic in the other

Student book references

2.23 Energy and chemical reactions

2.24 Energy in, energy out

Specification key

✔ C2.5.1

Energy transfer

When chemical reactions happen, energy is transferred to or from the surroundings. The energy may be transferred as heat, light, sound or movement.

Exothermic reactions

An **exothermic reaction** is one that transfers energy to the surroundings. Examples of exothermic reactions include

- combustion reactions
- many oxidation reactions
- neutralisation reactions.

Combustion (burning) reactions transfer energy to the surroundings, mainly as heat and light. The energy transferred in combustion reactions is useful for cooking, heating homes, and generating electricity.

The neutralisation reaction of dilute hydrochloric acid with sodium hydroxide solution is exothermic. When you mix the two solutions, the temperature increases as the reaction takes place. The temperature then decreases slowly to room temperature as energy is transferred as heat from the reaction mixture to the surroundings.

Substance	Temperature (°C)
sodium hydroxide solution, before mixing	20
hydrochloric acid, before mixing	20
reaction mixture, immediately after mixing	49
reaction mixture, 1 hour after mixing	20

reacting mixture of dilute hydrochloric acid and sodium hydroxide solution.

▲ When the reaction finishes, energy is transferred to the surroundings.

Using exothermic reactions

Exothermic reactions are used in items such as:

- Hand warmers – when you activate the hand warmer, there is an exothermic reaction. This transfers heat energy to your hands.
- Self-heating coffee cans – these have two compartments. One contains cold coffee. The other contains substances that react together in an exothermic reaction. The reaction transfers heat to the coffee.

Endothermic reactions

An **endothermic reaction** takes in energy from the surroundings.

Thermal decomposition reactions are endothermic. For example:

copper carbonate → copper oxide + carbon dioxide

$$CuCO_3 \rightarrow CuO + CO_2$$

Using endothermic reactions

Some sports injury packs are based on an endothermic reaction. When activated, there is a chemical reaction between the reactants in the pack. The temperature of the reacting mixture decreases, cooling the injury.

As the pack takes in heat from the injured muscle, the temperature of the pack gradually increases to the temperature of the surroundings.

Reversible reactions

If a reversible reaction is exothermic in one direction, it is endothermic in the opposite direction. The same amount of energy is transferred in each direction.

Blue copper sulfate crystals contain water – they are **hydrated**. Heating hydrated copper sulfate crystals forms white **anhydrous** copper sulfate powder, which contains no water. The process is endothermic.

Adding water to anhydrous copper sulfate powder is an exothermic process. Heat energy is transferred to the surroundings.

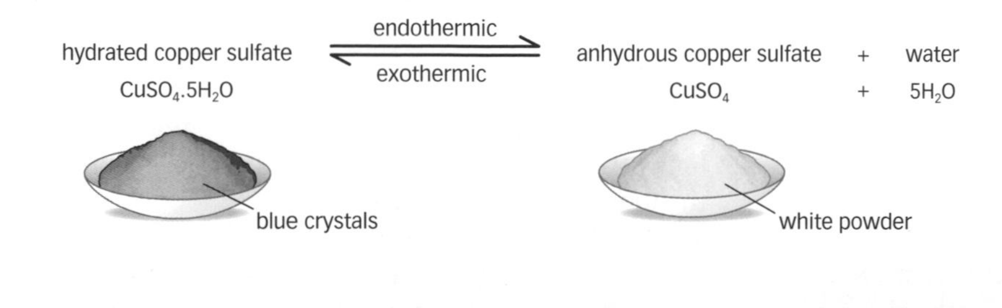

Key words

exothermic reaction, endothermic reaction, hydrated, anhydrous

Exam tip AQA

Remember – you go out of an **exit**, and **ex**othermic reactions give out energy.

Questions

1. What is an exothermic reaction?
2. Give an example of **one** way in which endothermic reactions can be useful.
3. Explain why, in an exothermic reaction, the temperature first increases, and then decreases back to the temperature of the surroundings.

Questions
Exothermic and endothermic reactions

Working to Grade E

1 Choose words from the box below to fill in the gaps in the sentences that follow. The words in the box may be used once, more than once, or not at all.

always never sometimes exothermic endothermic

In some chemical reactions, energy is transferred to the surroundings. These are ____________ reactions. Some reactions take in energy from the surroundings. These are ____________ reactions. Energy is ____________ transferred to or from the surroundings in chemical reactions.

2 Draw lines to match each type of reaction to one or more uses.

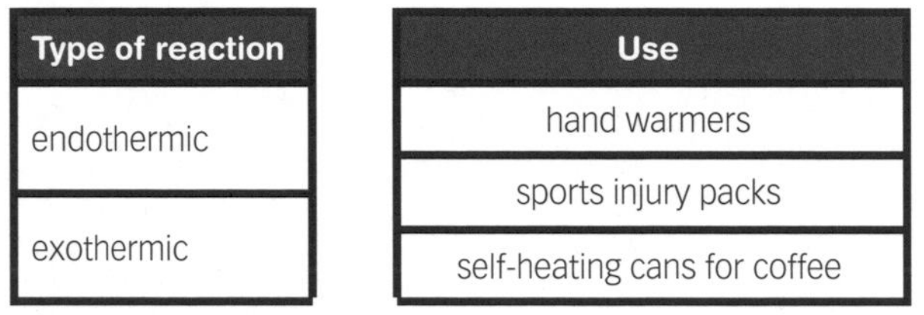

Type of reaction	Use
endothermic	hand warmers
exothermic	sports injury packs
	self-heating cans for coffee

3 Tick **one** box to show which type of reaction below is endothermic.

a Combustion ☐
b Oxidation ☐
c Neutralisation ☐
d Thermal decomposition ☐

4 A student neutralised an alkali with three different acids. She measured the temperature change in each neutralisation reaction.

Acid	Temperature change (°C)			
	Test 1	Test 2	Test 3	Mean
hydrochloric acid	25	26	27	26
nitric acid	20	5	20	20
ethanoic acid	14	9	10	

a Calculate the missing mean.
b Identify the anomalous result.
c Why did the student do three tests for each acid? Tick (✓) **two** answers.

Reason	Tick (✓)
To improve the resolution of the data	
To spot any anomalous data	
To spot any unreliable results	
To improve the accuracy of the data	

d The student wanted to make valid comparisons between the two acids. Which variables should she control? Tick the correct answers.

Variable	Tick (✓)
The volume of the acid	
The temperature of the acid	
The type of acid	
The concentration of the acid	

Working to Grade C

5 Put a tick next to each of the reactions below that is likely to be exothermic.

	Reaction	Tick (✓)
(a)	magnesium + oxygen → magnesium oxide	
(b)	copper carbonate → copper oxide + carbon dioxide	
(c)	methane + oxygen → carbon dioxide + water	
(d)	hydrogen + oxygen → water	
(e)	lithium nitrate → lithium oxide + nitrogen dioxide + oxygen	

6 A student mixed five pairs of solutions. On mixing, reactions occurred. The student measured the temperatures of the solutions before and immediately after mixing. His results are in the table.

Reaction	Temperature before reaction (°C)	Temperature immediately after mixing (°C)
A	20	54
B	20	71
C	20	5
D	20	82
E	20	14

An hour later, the student measured the temperatures of the reaction mixtures again. The temperature of each mixture was 20 °C, the same as the temperature of the room.

a Give the letters of the reaction mixtures that transfer energy to the surroundings as they return to room temperature.
b Give the letters of the reaction mixtures that take in energy as they return to room temperature.
c Give the letters of reactions that are endothermic.
d Give the letters of the reactions that are exothermic.

7 Annotate the equation below to show the direction in which the reaction is exothermic and the direction in which the reaction is endothermic.

$$\text{hydrated copper sulfate} \rightleftharpoons \text{anhydrous copper sulfate} + \text{water}$$

Exothermic and endothermic reactions

1 A student neutralised an acid with an alkali.

She recorded the temperatures of the solutions before and after the reaction.

Her results are in this table.

temperature of acid before reaction (ºC)	19
temperature of alkali before reaction (ºC)	21
maximum temperature reached after reaction (ºC)	69

a Explain how the results show that the neutralisation reaction is exothermic.

....

....

(1 mark)

b Which other types of reaction are usually exothermic?

Tick (✓) **two** boxes.

Type of reaction	Tick (✓)
thermal decomposition	
oxidation	
combustion	

(1 mark)

c Give an example of **one** way in which exothermic reactions are useful.

....

(1 mark)

d **i** Describe what happens to the temperature of a sports injury pack immediately after it is activated.

....

(1 mark)

ii Explain why the temperature of the sports injury pack gradually returns to that of the surroundings.

....

....

(1 mark)

(Total marks: 5)

Revision objectives

- explain what makes solutions acidic or alkaline
- explain what happens in neutralisation reactions

Student book references

2.25 Acids and bases

Specification key

- C2.6.1 a
- C2.6.2 a, c – e

State symbols

In symbol equations, **state symbols** give the states of the reactants and products:

- (s) means solid
- (l) means liquid
- (g) means gas
- (aq) means aqueous, or dissolved in water.

Bases, alkalis, and acids

Metal oxides and metal hydroxides are **bases**. Bases neutralise acids. Copper oxide, zinc oxide, and sodium hydroxide are examples of bases.

Some metal hydroxides are soluble in water. These hydroxides are **alkalis**. Sodium hydroxide and potassium hydroxide are examples of alkalis.

Hydroxide ions, OH^-(aq) make solutions alkaline.

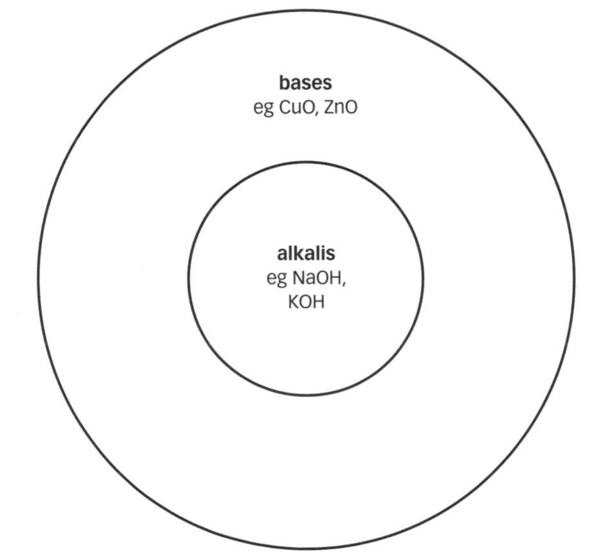

▲ All metal oxides and hydroxides are bases. Soluble hydroxides are called alkalis.

Ammonia

Ammonia, NH_3, dissolves in water to make an alkaline solution.

ammonia + water → ammonium hydroxide

$$NH_3(g) + H_2O(l) \rightarrow NH_4OH(aq)$$

In solution, the ammonium ions (NH_4^+) and the hydroxide ions (OH^-) are separated and surrounded by water molecules. The dissolved OH^- ions make the solution alkaline.

Compounds containing ammonium ions are important fertilisers.

pH scale

The **pH scale** is a measure of the acidity or alkalinity of a solution. On the pH scale:

- a solution of pH 7 is neutral
- a solution with a pH of less than 7 is acidic
- a solution with a pH of more than 7 is alkaline.

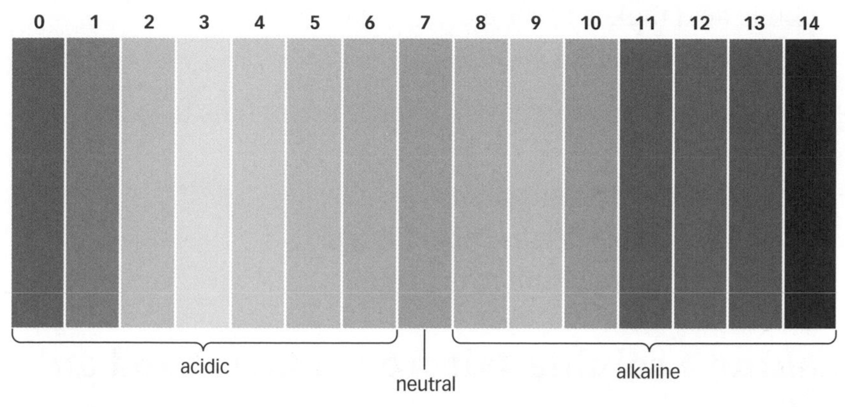

▲ The pH scale. Hydroxide ions (OH^-) make solutions alkaline and hydrogen ions (H^+) make solutions acidic. Universal indicator is usually red, orange, or yellow in acidic solutions, green in neutral solutions, and blue or purple in alkaline solutions.

Neutralisation reactions

You can use sodium hydroxide solution to neutralise hydrochloric acid. The products are sodium chloride and water:

hydrochloric acid	+	sodium hydroxide	→	sodium chloride	+	water
HCl(aq)	+	NaOH(aq)	→	NaCl(aq)	+	H_2O(l)

This is an example of a **neutralisation reaction**. In the reaction, OH^- ions from the sodium hydroxide react with H^+ ions from the acid to form water. You can represent neutralisation reactions like this:

$$H^+(aq) + OH^-(aq) \rightarrow H_2O(l)$$

Questions

1 Name **five** bases and **two** alkalis.

2 Explain why ammonia dissolves in water to make an alkaline solution.

3 Write a word equation for the neutralisation reaction of hydrochloric acid with potassium hydroxide.

4 **H** Write a symbol equation for the reaction of hydrochloric acid with potassium hydroxide. Include state symbols.

Key words

state symbol, base, alkali, hydroxide ion, pH scale, neutralisation reaction

Exam tip

AQA

Remember, an alkali is a soluble base. So all alkalis are bases, but not all bases are alkalis.

13: Making salts

Revision objectives

- explain what a salt is
- describe how to make soluble salts from acids and metals
- describe how to make soluble salts from acids and metal oxides
- describe how to make soluble salts from acids and alkalis

Student book references

2.26 Making soluble salts – 1

2.27 Making soluble salts – 2

Specification key

- C2.6.1 b – c
- C2.6.2 b

Salts

A **salt** is a compound that contains metal or ammonium ions. Salts can be made from acids. Different acids make different types of salt.

- Hydrochloric acid makes chlorides.
- Nitric acid makes nitrates.
- Sulfuric acid makes sulfates.

Several types of substance can supply metal ions to salts, including:

- metals, such as magnesium
- insoluble bases, such as copper oxide
- alkalis, such as sodium hydroxide.

Making a soluble salt from a metal and an acid

Soluble salts are salts that dissolve in water.

You can make some soluble salts by reacting a metal with an acid. For example, to make magnesium sulfate;

- keep adding small pieces of magnesium to sulfuric acid until there is no more bubbling and a little solid magnesium remains
- filter to remove unreacted magnesium
- heat the solution over a water bath, until about half its water has evaporated
- leave the solution to stand for a few days. Magnesium sulfate crystals will form. This is **crystallisation**.

The equation for the reaction is:

magnesium + sulfuric acid → magnesium sulfate + hydrogen

$$Mg(s) + H_2SO_4(aq) \rightarrow MgSO_4(aq) + H_2(g)$$

You can't make all metal salts like this. Some metals, such as copper, are not reactive enough. Some, like sodium, are too reactive – it is not safe to add sodium metal to dilute acids in a school science laboratory.

Exam tip

Remember – hydrochloric acid makes chloride salts, sulfuric acid makes sulfates, and nitric acid makes nitrates.

Making a soluble salt from an insoluble base and an acid

You can make some metal salts by reacting an insoluble base with an acid. For example, to make copper sulfate:

1. add copper oxide to sulfuric acid until some copper oxide remains unreacted
2. filter the mixture to remove the unreacted copper oxide
3. heat the solution over a water bath, until about half the water has evaporated
4. leave the solution to stand for a few days. Copper sulfate will crystallise from the solution.

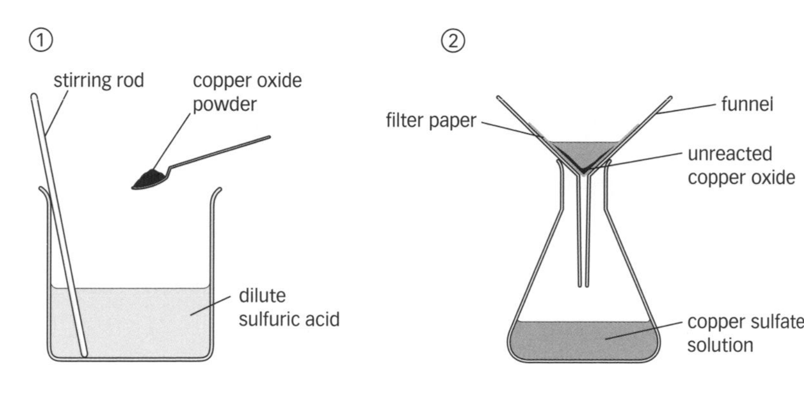

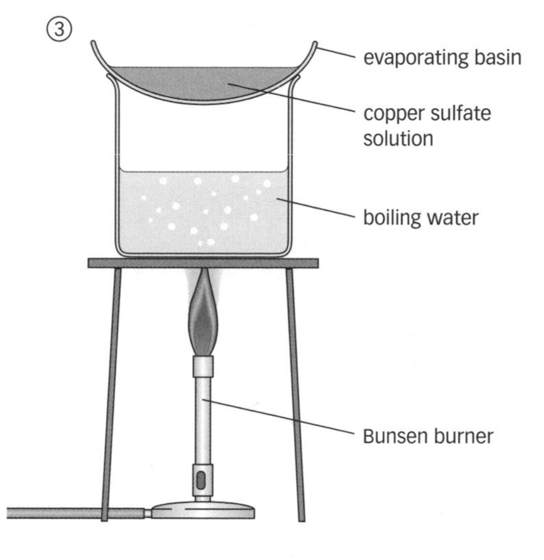

The equation for the reaction is:

copper oxide + sulfuric acid → copper sulfate + water

$$CuO(s) + H_2SO_4(aq) \rightarrow CuSO_4(aq) + H_2O(l)$$

Making a soluble salt from an acid and an alkali

The pictures below show how to make sodium chloride from an acid and an alkali. The equation for the reaction is:

sodium hydroxide + hydrochloric acid → sodium chloride + water

$$NaOH(aq) + HCl(aq) \rightarrow NaCl(aq) + H_2O(l)$$

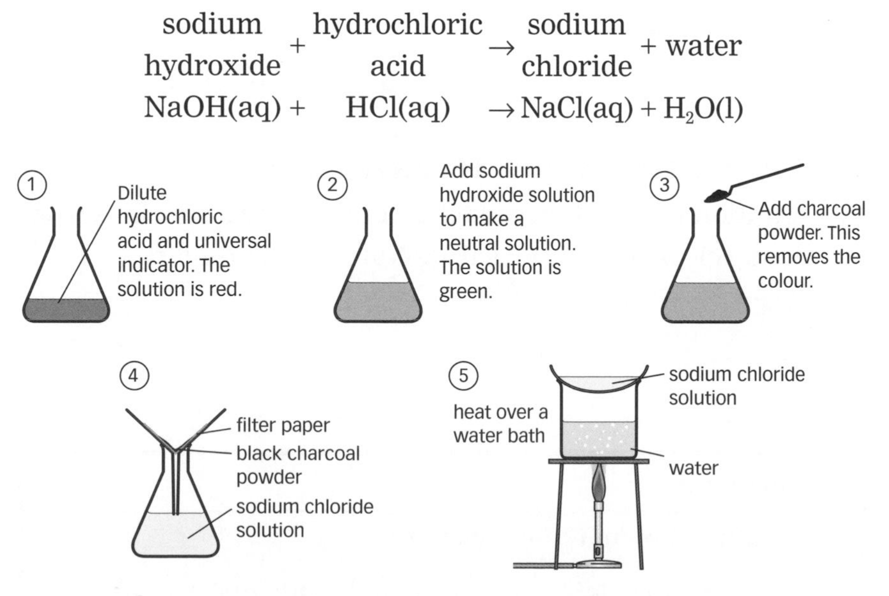

You can use different pairs of acids and alkalis to make different salts.

Key words

salt, soluble, crystallisation

Questions

1. What is a salt?
2. Describe how to make zinc sulfate from zinc and sulfuric acid.
3. Name the salt made by reacting zinc oxide with hydrochloric acid. Write a word equation for the reaction.
4. Describe how to make crystals of potassium nitrate from potassium hydroxide solution and an acid. Write a word equation for the reaction.

14: Precipitation and insoluble salts

Revision objectives

- predict the solutions needed to make an insoluble salt
- give an example of how precipitation reactions are useful

Student book references

2.28 Precipitation reactions

Specification key

- **C2.6.1 d**

Making insoluble salts

You can make insoluble salts from solutions in **precipitation** reactions. For example, reacting lead nitrate solution with potassium iodide solution produces a **precipitate** of lead iodide. A precipitate is a suspension of small solid particles, spread throughout a liquid or solution. It makes the liquid look cloudy.

$$\text{lead nitrate} + \text{potassium iodide} \rightarrow \text{lead iodide} + \text{potassium nitrate}$$

$$Pb(NO_3)_2(aq) + 2KI(aq) \rightarrow PbI_2(s) + 2KNO_3(aq)$$

The equation below shows only the ions that take part in the reaction. It is an **ionic equation**.

$$Pb^{2+}(aq) + 2I^-(aq) \rightarrow PbI_2(s)$$

The starting solutions are colourless. Lead iodide is bright yellow.

You can separate solid lead iodide from the potassium nitrate solution by filtering the mixture.

Predicting precipitates

To predict how to make insoluble salts, you need to know which salts are soluble in water.

Salts	Are they soluble?
nitrates	all soluble
chlorides	all soluble, expect for lead chloride and silver chloride
sulfates	all soluble, expect for lead sulfate, calcium sulfate, and barium sulfate

Lithium, sodium, and potassium salts are also soluble.

Using precipitation reactions

Precipitation reactions are used to remove unwanted ions from solutions, for example, in treating water for drinking.

Key words

precipitation, precipitate, ionic equation

Questions

1 What is a precipitation reaction?

2 Suggest **two** soluble salts you could use to make a precipitate of barium sulfate.

Questions
Acids, bases, and salts

Working to Grade E

1 Draw lines to match each state symbol to its meaning.

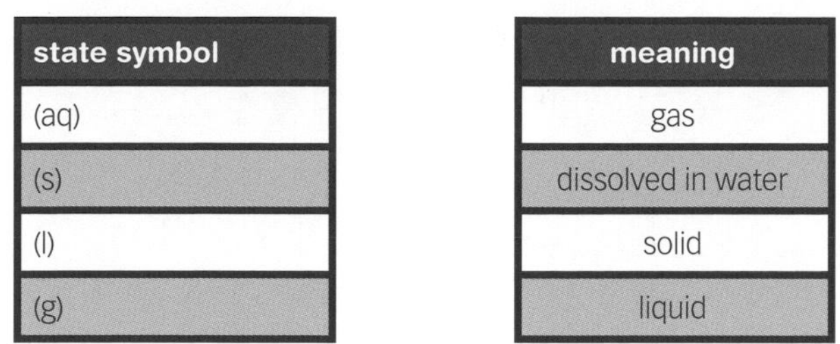

state symbol
(aq)
(s)
(l)
(g)

meaning
gas
dissolved in water
solid
liquid

2 In the sentences below, highlight the one bold word or phrase that is correct.
  a Hydrochloric acid makes **chlorides/ hydrochlorides**.
  b Nitric acid makes **nitrics/nitrates**.
  c Sulfuric acid makes **sulfides/sulfates**.

3 Choose words from the box below to fill in the gaps in the sentences that follow. The words in the box may be used once, more than once, or not at all.

ammonium acidic fertilisers sodium drinks ammonia alkaline

Ammonia dissolves in water to produce an ________ solution. It is used to produce ________ salts. These salts are important ________.

4 Highlight the statements below that are true. Then write corrected versions of the statements that are false.
  a Crystallisation is the formation of salt crystals from a salt solution.
  b Bases are oxides of non-metals.
  c Sulfur dioxide is a base.
  d Zinc oxide is a base.
  e Hydroxide ions make solutions alkaline.
  f Soluble hydroxides are called alkalis.
  g Potassium hydroxide is an alkali.
  h Magnesium oxide is a base.
  i Hydrogen ions make solutions alkaline.
  j The pH of a neutral solution is 7.
  k A solution with a pH of 6 is alkaline.

5 The statements below describe the steps in making a salt from magnesium and hydrochloric acid. Write the letters of the steps in the best order.
  A Filter the mixture.
  B Place the evaporating basin over a beaker of boiling water.
  C Add magnesium to the acid until some solid magnesium remains.
  D Leave in a warm place for a few days.
  E Pour the filtrate into an evaporating basin.
  F Heat until about half the water has evaporated from the solution.

Working to Grade C

6 Complete the table below.

Solution type	pH
acidic	________ than 7
neutral	
alkaline	________ than 7

7 Name the soluble salt made from each pair of substances below.
  a Magnesium and hydrochloric acid.
  b Copper oxide and sulfuric acid.
  c Potassium hydroxide and nitric acid.
  d Magnesium oxide and sulfuric acid.

8 Complete the neutralisation equation below.

$$H^+ \_\_\_ + \_\_\_ (aq) \rightarrow \_\_\_ (l)$$

9 State whether the solutions in the table are acidic, alkaline or neutral.

Solution contains...	Acidic, alkaline or neutral?
an equal number of OH^- and H^+ ions.	
more H^+ ions than OH^- ions.	
more OH^- ions than H^+ ions.	

10 Write instructions for making magnesium sulfate crystals from magnesium oxide and sulfuric acid.

11 Write instructions for making sodium chloride crystals from an acid and an alkali in a neutralisation reaction. Include the names of the starting materials.

12 Fill in the gaps in the sentences below to describe how to make solid lead iodide from two solutions.
Place a solution of lead ________ in a beaker. Add a solution of ________ iodide. A bright yellow ________ forms. Filter the mixture. Solid ________ ________ remains in the filter paper. The filtrate is a solution of ________ ________.

13 Name the insoluble salt made from each pair of solutions below:
  a lead nitrate and potassium iodide
  b barium chloride and sodium sulfate
  c lead nitrate and sodium iodide
  d silver nitrate and sodium hydroxide
  e barium nitrate and lithium sulfate.

14 Suggest pairs of solutions that could be used to make the insoluble salts listed below:
  a lead chloride
  b calcium sulfate
  c lead sulfate
  d barium sulfate.

Examination questions
Acids, bases, and salts

1 A student wanted to make zinc chloride crystals by reacting a metal with an acid.

a Which acid should the student use?

Tick (✓) **one** box.

Acid	Tick (√)
hydrochloric acid	
nitric acid	
sulfuric acid	

(1 mark)

b The student placed some acid in a conical flask.

He added zinc metal until no more would react.

He filtered the resulting mixture.

Explain why the student filtered the mixture.

....

....

(1 mark)

c The student poured the zinc chloride solution into an evaporating dish.

He heated it over a water bath until half the solution had evaporated.

He placed the dish and its contents in a warm place.

A week later, the student observed crystals in the dish.

Name the process by which zinc chloride crystals formed from the zinc chloride solution.

Tick (✓) **one** box.

Name of process	Tick (✓)
filtration	
neutralisation	
solution	
crystallisation	

(1 mark)

d The hazard symbols below appear on a bottle of zinc chloride crystals.

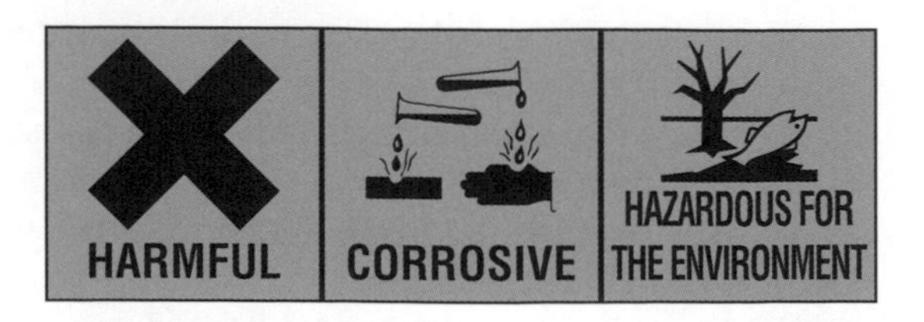

Suggest **two** ways of managing the risks from these hazards when dealing with the zinc chloride made in the experiment.

1 ....

2 ....

(2 marks)

(Total marks: 5)

2 The table gives data about four oxides.

Name of oxide	Does it dissolve in water?
copper oxide	no
magnesium oxide	no
sodium hydroxide	yes
carbon dioxide	yes

a Use data from the periodic table and the table above to name:

i **three** bases

....

(1 mark)

ii **one** alkali

....

(1 mark)

b Give the name and formula of the ion that makes solutions alkaline.

Name ....

Formula ....

(2 marks)

(Total marks: 4)

3 A student plans to make solid lead iodide in a precipitation reaction.

a Name **two** solutions that the student could use.

1 ....

2 ....

(2 marks)

b Describe how the student could make solid lead iodide from the two solutions you named in part a.

....

....

(2 marks)

c The equation below shows the ions that take part in the precipitation reaction to make lead iodide.

Complete the equation by adding state symbols for each substance shown in the equation.

Pb^{2+}......... + 2 I^- .... → PbI_2 ....

(3 marks)

(Total marks: 7)

15: Electrolysis 1

Revision objectives

- describe what happens at the electrodes in electrolysis
- use half equations to represent reactions at electrodes

Student book references

2.29 Electrolysis

Specification key

- C2.7.1 a – c, e – g

What is electrolysis?

When an ionic substance is molten, or dissolved in water, its ions can move about.

Passing an electric current through the liquid or solution breaks down the ionic compound into simpler substances. This is **electrolysis**. The substance that is broken down is the **electrolyte**.

During electrolysis, the ions move to the **electrodes**:

- Positive ions move to the negative electrode.
- Negative ions move to the positive electrode.

This diagram shows what happens when an electric current passes through molten lead bromide.

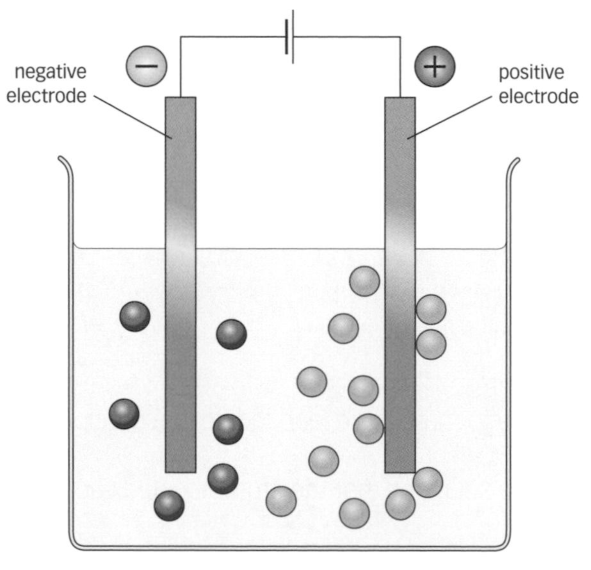

▲ The electrolysis of molten lead bromide.

In the lead bromide electrolysis cell two things happen:

- Positively charged lead ions move towards the negative electrode. Lead metal forms here.
- Negatively charged bromide ions move towards the positive electrode. Liquid bromine forms here.

What happens at the electrodes?

- At the negative electrode, positively charged ions gain electrons. This is **reduction**. In the example, lead ions gain electrons. The positive ions are reduced.
- At the positive electrode, negatively charged ions lose electrons. This is **oxidation**. The negative ions are oxidised.

H Half equations

Chemists use **half equations** to show what happens at the electrodes. In half equations, an electron is represented as e^-. You can balance half equations by adding or subtracting electrons until the total charge on each side is equal.

For example, in the electrolysis of lead bromide:

- at the negative electrode, $Pb^{2+} + 2e^- \rightarrow Pb$
- at the positive electrode, $Br^- \rightarrow Br + e^-$

This is also written as $Br^- - e^- \rightarrow Br$

The bromine atoms then join together in pairs to make bromine molecules:

$$Br + Br \rightarrow Br_2$$

Key words

electrolysis, electrolyte, electrode, reduction, oxidation, half equation

Predicting products

If you electrolyse molten lead bromide, there is only one possible product at each electrode. But for compounds that are dissolved in water, there are other possible products, since water also takes part in electrolysis reactions.

The table below shows the products when electricity passes through some solutions.

Solution	Product at negative electrode	Product at positive electrode
copper chloride	copper	chlorine
magnesium iodide	hydrogen	iodine
silver nitrate	silver	oxygen
sodium sulfate	hydrogen	oxygen
potassium carbonate	hydrogen	oxygen

At the negative electrode:

- the metal is produced if it is low in the reactivity series
- hydrogen is produced if the metal is above copper in the reactivity series. The hydrogen is formed by the electrolysis of water.

At the positive electrode:

- halogens are produced if there are halide ions in the solution
- oxygen is produced if there are nitrate, sulfate or carbonate ions in the solution. The oxygen comes from the water.

Exam tip AQA

Remember OIL RIG – Oxidation Is Loss (of electrons) and Reduction Is Gain (of electrons).

Questions

1 Name the products formed at the positive and negative electrodes in the electrolysis of molten copper chloride.

2 Predict the products formed at the electrodes during the electrolysis of a solution of sodium carbonate in water.

3 **H** Write half equations for the reactions that occur at the electrodes during the electrolysis of molten copper chloride.

16: Electrolysis 2

Revision objectives

- describe how electroplating works
- describe and explain how aluminium is extracted from aluminium oxide by electrolysis
- name the products of the electrolysis of sodium chloride solution

Student book references

2.30 Electroplating
2.31 Using electrolysis – 1
2.32 Using electrolysis – 2

Specification key

✔ C2.7.1 d, h – i

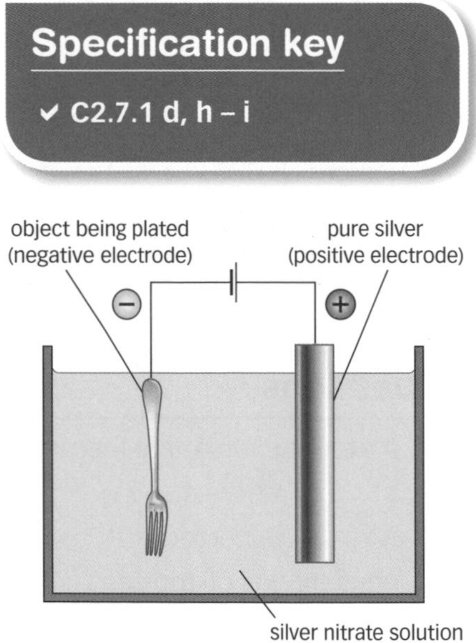

▲ This diagram shows how to electroplate an object with silver.

Electroplating

Electrolysis is used to coat objects with a thin layer of a metal such as copper, silver or tin. This is **electroplating**. Electroplating has the following two main purposes:

- to protect a metal object from corrosion by coating it with an unreactive metal that does not easily corrode
- to make an object look attractive.

Food cans are made of tinplate – steel that has been electroplated with tin. Cutlery may be electroplated with silver.

Extracting aluminium

Aluminium is extracted from **bauxite ore**. Bauxite is mainly aluminium oxide. Pure aluminium is obtained from aluminium oxide in the following way.

Dissolve aluminium oxide in molten **cryolite**. The solution formed has a lower melting point than pure aluminium oxide. So less energy is needed to keep the mixture liquid than would be needed for aluminium oxide alone.

↓

Pour the aluminium oxide and cryolite mixture into a huge electrolysis cell.

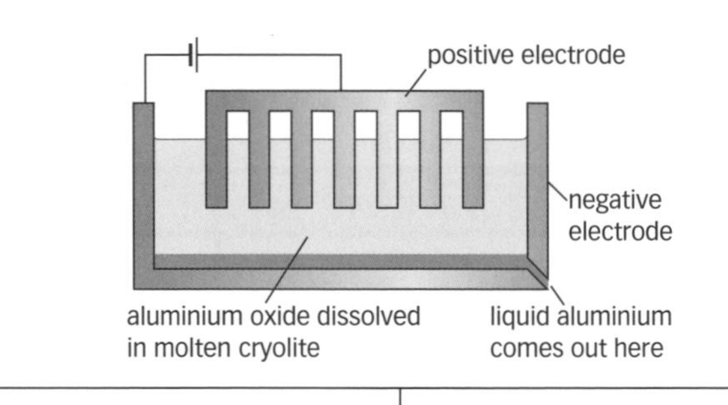

↓

Pass an electric current through the liquid mixture.

↓

Aluminium ions move to the negative electrode. They gain electrons to make liquid aluminium metal.

H $Al^{3+} + 3e^- \rightarrow Al$

↓

Oxide ions move to the positive electrode. They lose electrons to form oxygen atoms, which join together in pairs to make oxygen gas. The oxygen gas reacts with the carbon of the electrode and forms carbon dioxide gas.

H $O^{2-} \rightarrow O + 2e^-$ then $O + O \rightarrow O_2$ then $C + O_2 \rightarrow CO_2$

The electrolysis of sodium chloride solution

The electrolysis of concentrated sodium chloride solution, or **brine**, produces several products:

- Hydrogen gas forms at the negative electrode.
- Chlorine gas forms at the positive electrode.

A solution of sodium hydroxide also forms during the process.

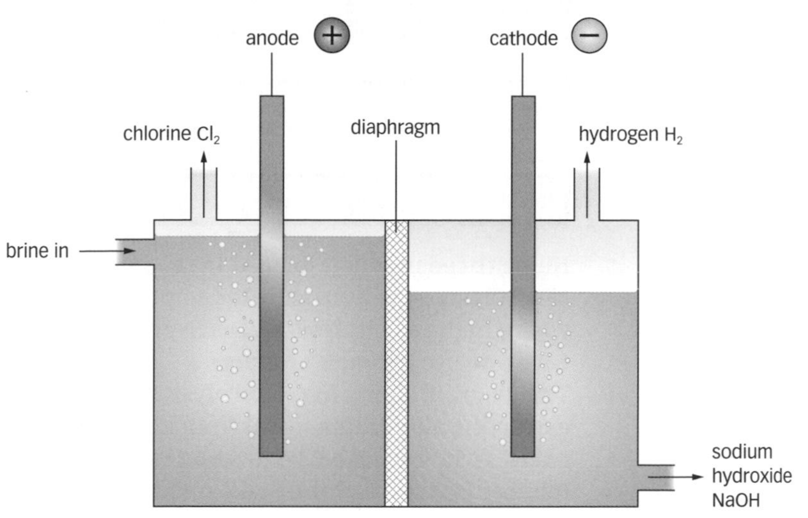

▲ The apparatus used in the industrial electrolysis of brine.

The products have many uses:

- Sodium hydroxide is used to make soap.
- Chlorine is used to make bleach and plastics, and to sterilise water.
- Hydrogen is used to make margarine and ammonia.

Key words

electroplating, bauxite ore, cryolite, brine

Exam tip

AQA

Remember, in electroplating the object needs to be placed at the negative electrode.

Questions

1. Give **two** reasons for electroplating objects.
2. Name the **two** products formed in the industrial electrolysis of aluminium oxide.
3. Name the substances formed at each electrode in the electrolysis of sodium chloride solution.

Questions
Electrolysis

Working to Grade E

1 Tick **two** possible reasons for electroplating an object.

a To protect the object from corrosion. ☐
b To make the object look more attractive. ☐
c To make the object harder to detect with a metal detector. ☐

2 Choose words from the box below to fill in the gaps in the sentences that follow. The words in the box may be used once, more than once, or not at all.

solid evaporated melted atoms ions solution dissolved

When an ionic substance is _________ or _________ in water, the _________ are free to move about within the liquid or _________.

3 Draw lines to match each word to its definition.

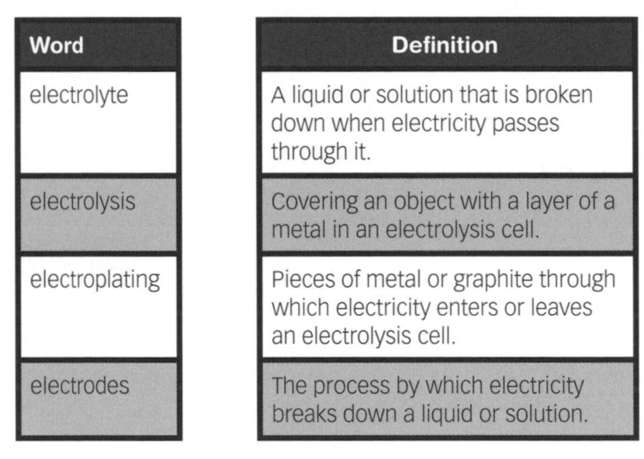

Word	Definition
electrolyte	A liquid or solution that is broken down when electricity passes through it.
electrolysis	Covering an object with a layer of a metal in an electrolysis cell.
electroplating	Pieces of metal or graphite through which electricity enters or leaves an electrolysis cell.
electrodes	The process by which electricity breaks down a liquid or solution.

4 Highlight the correct word in each pair of bold words.
Molten lead bromide is made up of negatively charged **bromide/bromine** ions and **negatively/positively** charged lead ions. A student passes electricity through molten lead bromide. The negative ions move to the **positive/negative** electrode. The lead ions move to the **positive/negative** electrode.

Working to Grade C

5 Highlight the statements below that are true. Then write corrected versions of the sentences that are false.

a At the negative electrode, positively charged ions lose electrons.
b At the positive electrode, negatively charged ions lose electrons.
c If an ion gains electrons, the ion is oxidised.
d Reduction occurs when an ion gains electrons.

6 Predict what is formed at the negative and positive electrodes when electricity passes through the solutions in the table.

Solution	Positive electrode	Negative electrode
copper chloride		
potassium bromide		
silver nitrate		
magnesium nitrate		
copper carbonate		
sodium sulfate		

7 Name **three** products formed in the industrial electrolysis of sodium chloride solution. State where in the electrolysis cell each one is produced. Give **one** use for **each** product.

8 The diagram shows the electrolysis cell for extracting aluminium metal from aluminium oxide. Write each label below next to the correct number on the diagram. Some numbers on the diagram have more than one label.

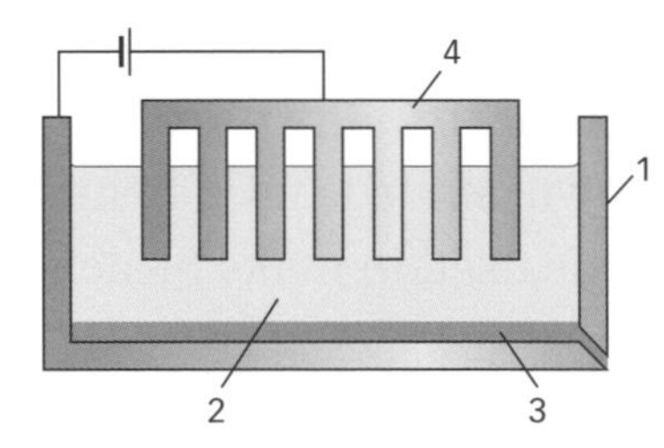

Labels:

A Positive electrode
B Positive ions gain electrons at this electrode
C Negative electrode
D Electrolyte of liquid aluminium and cryolite
E Liquid aluminium forms at this electrode
F Oxide ions are oxidised at this electrode
G Carbon dioxide gas forms here
H This electrode is made of carbon
I Reduction happens at this electrode
J Negative ions are attracted to this electrode

Working to Grade A*

9 Write half equations for the reactions that happen at the positive and negative electrodes during the electrolysis of:

a Molten lead bromide, $PbBr_2$
b Copper chloride solution, $CuCl_2$
c Molten aluminium oxide, Al_2O_3

Examination questions
Electrolysis

1 Some silver jewellery is electroplated with a layer of rhodium.

a The table gives some data for silver and rhodium.

	silver	**rhodium**
relative hardness	25	100
melting point (°C)	961	1970
colour	silvery-white	silvery-white
corrosion	reacts with hydrogen sulfide in the air to make black silver sulfide	does not react with gases in the air

Use the data in the table to help you suggest two reasons for electroplating silver jewellery.

1 ....

2 ....

(2 marks)

b The diagram below shows an electrolysis cell used to electroplate a silver ring.

The electrolyte contains positively charged rhodium ions, Rh^{3+}.

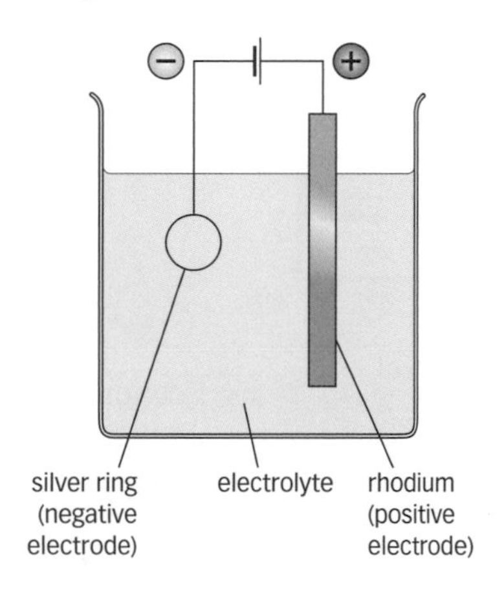

i The silver ring is used as the negative electrode in the electrolysis cell. Explain why.

....

(1 mark)

ii At the negative electrode, rhodium ions gain electrons to become rhodium atoms.

What type of reaction is this?

Put a ring around the one correct answer.

corrosion **oxidation** **reduction** **electrolysis** *(1 mark)*

H

iii Write a half equation to represent the reaction that happens at the negative electrode.

....

(1 mark)

(Total marks: 5)

Designing an investigation, presenting data, and using data to draw conclusions

Skill – Understanding the experiment

1 A student investigates the hypothesis *There is a relationship between the rate of a reaction and temperature.* She sets up the apparatus below.

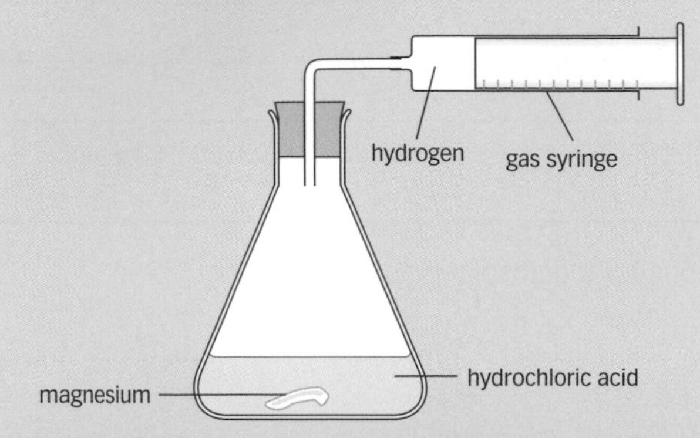

The student's results are in the table.

Temp (°C)	Time to collect 50 cm³ of hydrogen gas (s)			
	Test 1	Test 2	Test 3	Mean
20	401	401	404	402
30	196	197	195	196
40	376	100	100	192
50	47	47	50	
60	25	25	25	25

a Identify the dependent variable, the independent variable and **three** control variables for the investigation.

To answer this question, you need to know that the independent variable is the one that the student changes. The dependent variable is the one that the student measures for each change in the independent variable.

The control variables are the variables that the student keeps the same – these are not given in the question. You will need to use your knowledge and experience of doing investigations, and the diagram, to help you answer this part of the question.

Skill – Using data to draw conclusions

b Calculate the missing mean.

The mean is the sum of the measurements divided by the number of measurements taken. For example, the mean of the time values at 20 °C is:

$(401\,s + 401\,s + 404\,s) \div 3 = 402\,s$

c Plot the results on a graph.

Both the independent and the dependent variables are continuous – they can take any numerical value. This means that you can draw a line graph to display the data. Don't forget to draw a line of best fit rather than joining up all the points.

d Do the results support the hypothesis? Explain why.

To answer this question you will need to look back at the hypothesis at the start of the question. If the line graph you plotted in part c shows the relationship given in the hypothesis, then the results support the hypothesis.

Societal aspects of scientific evidence

Skill – Analysing the facts and making deductions

2 Making ibuprofen

In the past, an aluminium chloride catalyst was used in the manufacture of the painkiller ibuprofen. This catalyst could not be separated from the reaction mixture, and so could not be reused.

Today, two other catalysts are used in the production of ibuprofen – hydrogen fluoride and a nickel/aluminium alloy. These catalysts can be used again and again.

a Suggest an economic benefit of reusing a catalyst, compared to using a fresh sample of catalyst for each batch.

Economic factors are those to do with money, so you will need to explain why reusing a catalyst might help to increase the profits of a company making ibuprofen.

Skill – Understanding the impact of a decision

b Suggest **two** environmental benefits of reusing a catalyst.

Environmental factors might be linked to energy requirements (for example, those to extract raw materials, or to manufacture a product). Environmental factors are also linked to waste disposal. For example, how are unwanted by-products of an industrial process disposed of? What happens to a product once it has done its job, and is no longer useful?

Answering an extended writing question

QUESTION

1 *In this question you will be assessed on using good English, organising information clearly, and using specialist terms where appropriate.*
Hydrochloric acid reacts with calcium carbonate to make calcium chloride, water, and carbon dioxide gas. Describe and explain the factors that affect the rate of this reaction. *(6 marks)*

if u like make it more conc then it is faster becoz like the particals hit each other lots more often and there is a gass so the preshure going up makes it fasster and if you make it collder it shoud speed up I reckon

G–E

Examiner: This answer is typical of a grade-G candidate. It is worth just one mark, gained for describing and explaining that increasing the acid concentration increases the reaction rate. The candidate is incorrect in stating that increasing the pressure increases the rate.

The candidate has not used specialist terms. There is no punctuation, and there are several spelling mistakes.

If temp go up so does the rate becuase the particals move faster.
also if you have powder with big surface area it is faster, and can bubble up right to the top of the flask and even overflow all over everywhere and make a mess very quickly and the teacher might get annoyed because it is dangerous. With dilute acid it is slower than with consentrated acid.

D–C

Examiner: This answer is worth three marks out of six. It is typical of a grade-C or -D candidate. The candidate has described three factors that affect reaction rate, and partially explained one of these in terms of particles and collision theory.

The answer is well organised, with a few mistakes of grammar and punctuation. There are spelling mistakes. Part of the answer is not relevant.

The rate can be affected by acid concentration, temperature and surface area of the calcium carbonate.
Increasing the acid concentration increases the rate. This is because there are more acid particles in a certain volume of solution, so they can collide more frequently with the calcium carbonate.
Increasing surface area of the calcium carbonate increases the rate because with a powder there are more particles exposed and ready to react. So the collisions are more frequent and the rate goes up.
Increasing temperature means the particles move faster. So they collide more often. They also collide with more energy. This increases the rate.

B–A*

Examiner: This is a high-quality answer, typical of an A* candidate. It is worth six marks out of six.

The question has been answered accurately and the explanations are detailed.

The answer is well organised, and includes a short introduction. The spelling, punctuation, and grammar are faultless. The candidate has used several specialist terms.

2 *In this question you will be assessed on using good English, organising information clearly, and using specialist terms where appropriate.*

Describe how to make copper sulfate crystals from an insoluble metal oxide and a dilute acid. *(6 marks)*

QUESTION

Mix up hydraulic acid and Copper Oxsyde. Heat it all up? Stir and filter! You've got xtals.

G–E

Examiner: This answer is worth one mark out of six, for correctly identifying the starting materials. It is typical of a grade-F or -G candidate.

There are several spelling and punctuation mistakes, and the stages described are not in the correct order.

Add copper oxide powder to hydrochloric acid. Stir. You adds a bit more. Then you filters it all. Keep the blue liquid. Heat it up and you made crystals round the edge of the dish.

D–C

Examiner: This answer is worth three marks out of six, and is typical of a grade-C or -D candidate. The candidate has given one incorrect starting material, and has described the stages in insufficient detail. The candidate has not described the final stage of the preparation of the crystals.

The answer is well organised. The spelling and punctuation are accurate, but there are some grammatical errors.

Place some dilute sulfuric acid in a beaker. Add copper oxide powder with a spatula, and stir, until no more will dissolve. Then filter to remove the unreacted black solid from the blue solution. Pour the blue solution – the filtrate – into an evaporating basin. Heat over a water bath until half the water has evaporated. Now the solution is more concentrated.
Then leave the solution in a warm place for a week. Crystals will form.

B–A*

Examiner: This answer gains six marks out of six, and is typical of an A or A* candidate. It describes in detail, using scientific words and correct apparatus names, how to make the salt from correctly identified starting materials.

The answer is logically organised, and includes scientific words. The spelling, punctuation and grammar are accurate.

Part 2 Course catch-up
Rates, energy, salts, and electrolysis

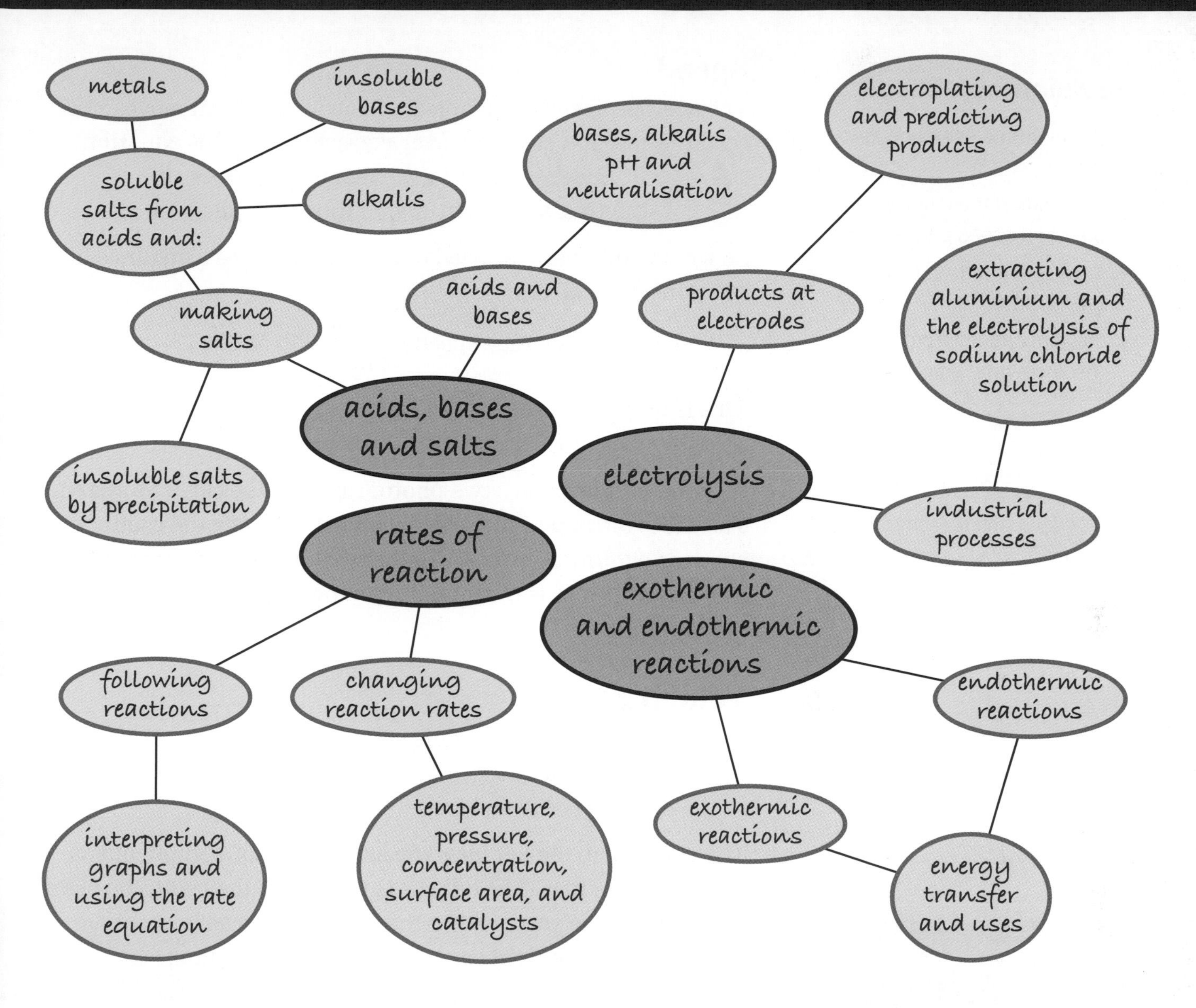

Revision checklist

- You can find the rate of a chemical reaction by measuring the amount of a product used or the amount of a product formed over time.
- Rate of reaction $= \frac{\text{amount of reactant used}}{\text{time}}$ or $\frac{\text{amount of product made}}{\text{time}}$
- Chemical reactions only happen when reacting particles collide with each other and with enough energy.
- Increase the rate of a reaction by increasing temperature, or pressure (of reacting gases), or concentration (of reacting solutions), or surface area (of reacting solids) or by adding a catalyst.
- Exothermic reactions transfer energy to the surroundings.
- Endothermic reactions take in energy from the surroundings.
- Soluble salts are made from acids by reacting them with metals, or insoluble bases, or alkalis. The salt solutions can be crystallised to make solid salts.
- Different acids produce different salts – hydrochloric acid makes chlorides, nitric acid makes nitrates, sulfuric acid makes sulfates.
- Insoluble salts are made in precipitation reactions.
- Hydrogen ions (H^+) make solutions acidic and hydroxide ions (OH^-) make solutions alkaline. These ions react together in neutralisation reactions.
- Ionic substances conduct electricity when liquid or dissolved in water since their ions are free to move.
- In electrolysis, positive ions are attracted to the negative electrode. Here they gain electrons. This is reduction.
- Negative ions are attracted to the positive electrode. Here they lose electrons. This is oxidation.
- Aluminium is extracted from aluminium oxide by the electrolysis of a molten mixture of aluminium oxide and cryolite. Aluminium forms at the negative electrode. Carbon dioxide forms at the positive electrode, which is made of carbon.
- The electrolysis of sodium chloride solution produces hydrogen at the negative electrode and chlorine at the positive electrode. Sodium hydroxide solution is also formed.

1: Speed and velocity

Revision objectives

- calculate the speed of an object
- understand the difference between speed and velocity
- draw and interpret distance–time graphs
- calculate speed from a distance–time graph

Student book references

2.3 Speed and velocity

2.4 Distance–time graphs

Specification key

- P2.1.2 b – d

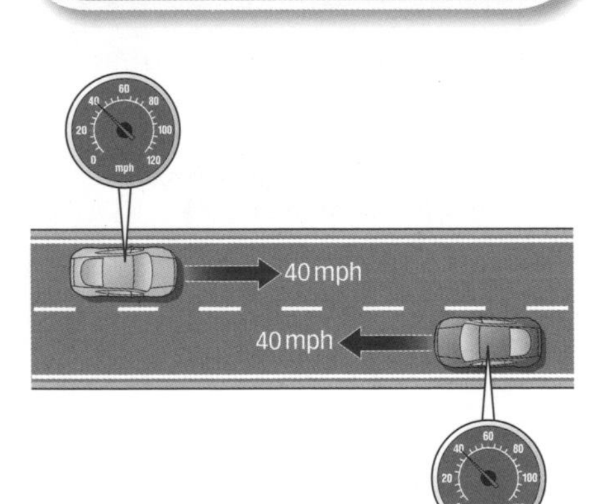

▲ Two cars with the same speed but a different velocity.

Speed

The average **speed** of an object tells you how far it moves in a certain time. You calculate average speed with this equation:

$$\text{average speed (metres/second, m/s)} = \frac{\text{distance (metres, m)}}{\text{time (seconds, s)}}$$

So what is the average speed of a car that moves a distance of 180 m in a time of 12 s?

$$\text{average speed} = \frac{180\,\text{m}}{12\,\text{s}} = 15\,\text{m/s}$$

This is only an average speed because the car may be speeding up or slowing down at the time of measurement.

A **speed camera** takes two photographs of the car, 0.5 s apart. The car is moving at 20 m/s, so its position along the road changes by $20\,\text{m/s} \times 0.5\,\text{s} = 10\,\text{m}$ between the two photographs. The faster the car is moving, the further it goes between the photographs.

Velocity

The **velocity** of an object tells you two things:

- how fast it is moving
- the direction of its motion.

The two cars to the left have the same speed of 40 mph, but different velocities. The left car is moving to the right, the right car to the left. If you say that the left car has a velocity of +40 mph, then the right car has a velocity of –40 mph.

Distance–time graphs

Time (s)	Laps	Distance (m)
0	0	0
20	1	300
40	2	600

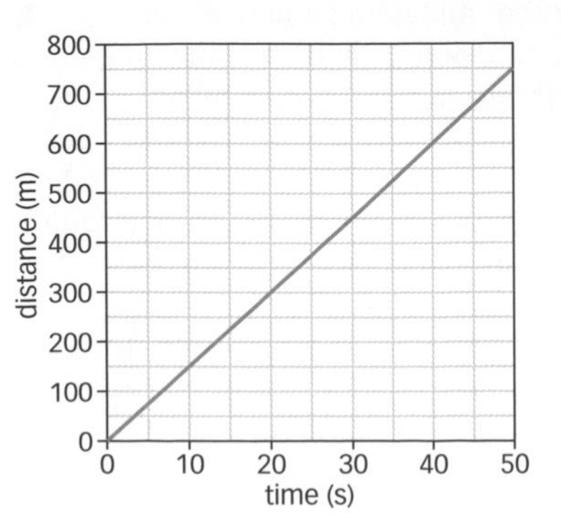

▲ Distance–time graph for a car using the data in the table.

A car is moving round and round a track at a steady speed. It has to move a distance of 300 m to get back to where it started. The table shows the reading on a stopwatch for two laps of the circuit. This information is presented as a **distance–time graph**, with time plotted on the *x*-axis. The car is going at a steady speed, so the graph is a straight line.

Key words

speed, speed camera, velocity, distance–time graph, gradient

H Gradient

Here is how to use the **gradient** (or slope) of the graph to calculate the speed of the car.

- Take any pair of data points (*x*, *y*) from the graph: (10 s, 150 m) and (50 s, 750 m)
- Calculate the increase of distance: 750 m – 150 m = 600 m
- Calculate the increase in time: 50 s – 10 s = 40 s
- Calculate the gradient:

$$\text{gradient (speed)} = \frac{\text{increase of distance}}{\text{increase in time}}$$

$$= \frac{600\,\text{m}}{40\,\text{s}}$$

$$= 15\,\text{m/s}$$

Questions

1 Joe runs a distance of 1200 m in 240 s. Calculate his average speed.

2 The three graphs below show how the distance of three cars from you changes with time.

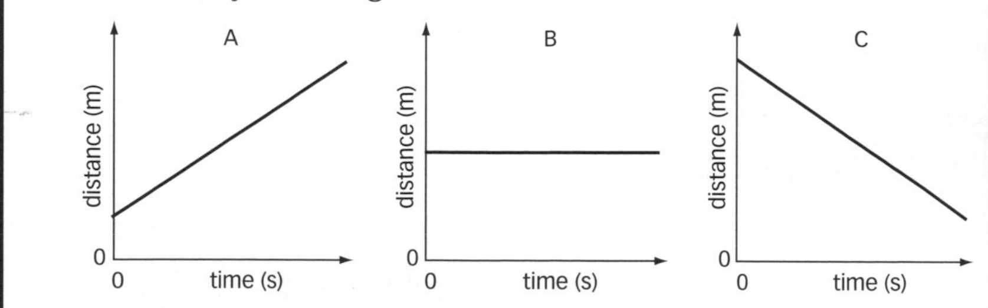

a Which car is not moving?

b Which car is moving towards you?

3 **H** Calculate the speed of runner P from the graph below.

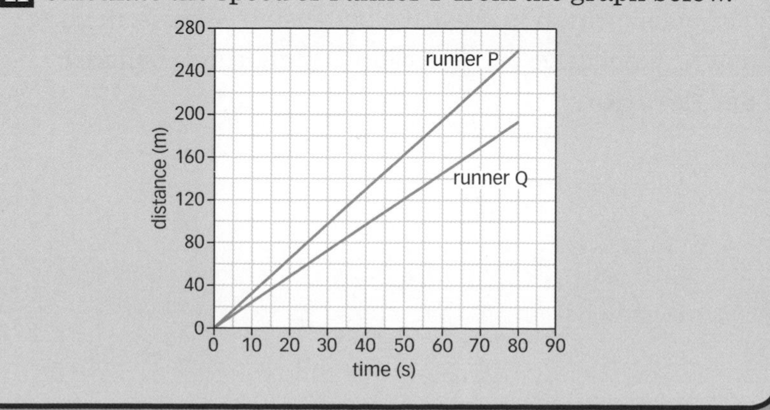

Exam tip

AQA

Practise using the speed equation to calculate distance or time.

Revision objectives

- ✔ calculate the acceleration of an object
- ✔ draw and interpret velocity–time graphs
- ✔ use a velocity–time graph to calculate acceleration or distance

Student book references

Specification key

✔ P2.1.2 e – h

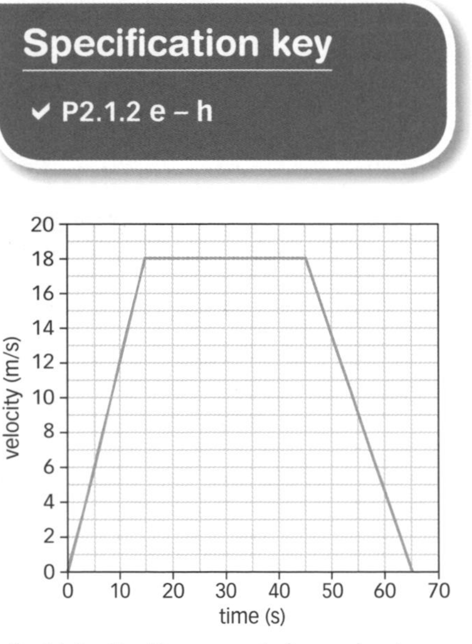

▲ Velocity–time graph for a short car journey.

Changing speed

Below is the **velocity–time graph** for a car. This is what it shows:

- The speed increases from zero in the first 15 s, so the car is accelerating.
- The speed is steady at 18 m/s for the next 30 s.
- The speed drops back to zero in the last 20 s, so the car is decelerating.

Calculating acceleration

The **acceleration** of an object tells you how fast its velocity is changing. You calculate acceleration with this equation:

$$a = \frac{v - u}{t}$$

a is the acceleration in metres per second squared (m/s^2)
u is the initial velocity in metres per second (m/s)
v is the final velocity in metres per second (m/s)
t is the time taken for the change in seconds (s)

So what is the acceleration of the car in the first 15 s of its journey?

The graph gives $u = 0\,m/s$, $v = 18\,m/s$, and $t = 15\,s$

$$a = \frac{v - u}{t} = \frac{18\,m/s - 0\,m/s}{15\,s} = 1.2\,m/s^2$$

This means that the speed of the car increases by 1.2 m/s in every second.

H Using gradients

What is the acceleration of the car as it slows down?

The graph gives $u = 18\,m/s$ at a time of 45 s and $v = 0\,m/s$ at a time of 65 s.

$$a = \frac{v - u}{t} = \frac{0\,m/s - 18\,m/s}{65\,s - 45\,s} = \frac{-18\,m/s}{20\,s} = -0.9\,m/s^2$$

The acceleration is negative because the car is slowing down. A negative acceleration is sometimes called a **deceleration**.

Backwards and forwards

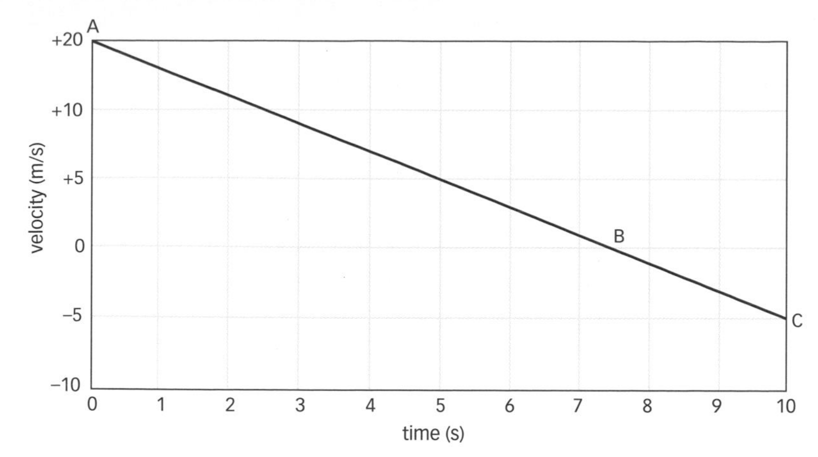

What does the above velocity–time graph for a car show?

- At point A, the car moves forwards with a velocity of +20 m/s.
- The velocity drops steadily to zero in the next 7.5 s as the car slows to a halt at point B.
- For the final 2.5 s the velocity is negative. This means that the car is now going backwards, ending up with a velocity of –5 m/s at point C.

H Distance

You calculate the distance moved by the car during this manoeuvre by finding the area between the line and the time axis of the velocity–time graph.

Remember the area of a triangle is ½ × base × height.

- The area of the triangle under the line AB = ½ × 7.5 s × 20 m/s = 75 m
- The area of the triangle under the line BC = ½ × 2.5 s × –5 m/s = –6.25 m

So the car moved forwards by 75 m, then moved backwards by 6.25 m. The overall distance covered was therefore 75 – 6.25 = 68.75 m.

Key words

velocity–time graph, acceleration, deceleration

Exam tip AQA

Always use changes of velocity and time when calculating acceleration.

Practise changing $a = \frac{v-u}{t}$ to $a \times t = v - u$ and $t = \frac{v-u}{a}$

Questions

1 The velocity of an accelerating car changes from 2 m/s to 17 m/s in 10 s. Calculate the acceleration of the car.

2 a Calculate the acceleration of car D in the graph below.

b How can you tell from the graph that car E has a greater acceleration?

3 **H** Calculate the distance that car D moves as the time increases from 0 s to 8 s.

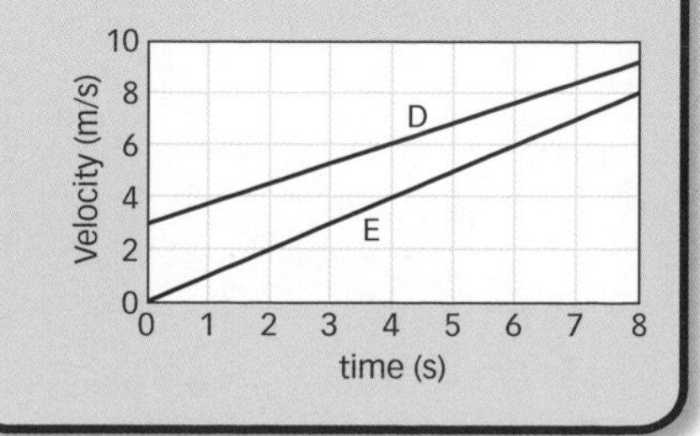

3: Force and acceleration

Revision objectives

- understand how forces change the motion of objects
- understand that forces between objects are equal and opposite to each other
- calculate the resultant force on an object
- use the equation *resultant force = mass × acceleration* ($F = m \times a$)

Student book references

2.1 Forces

2.2 Force, mass, and acceleration

Specification key

✔ P2.1.1 ✔ P2.1.2 a

▲ Forces always come in pairs.

Forces

A **force** is a push or a pull that changes the shape or motion of an object. All forces have a size (measured in newtons, N) and a direction. Here are some examples:

- Gravity exerts a downwards force of 650 N on a teenager.
- Friction exerts a backwards force of 6000 N on a braking car.
- The ground underneath an elephant exerts an upwards force of 10 000 N on each foot.

Pairs of forces

Every force is the result of one object interacting with another object. So the boy in the diagram interacts with the wall by pushing it to the right with a force of 100 N. The wall reacts by pushing back on the boy with another force of the same size (100 N), but in the opposite direction (left). Each force in a **pair of forces** between different objects has the same size but opposite direction.

Resultant force

When more than one force acts on an object you can combine them to find the **resultant force**. You need to take account of the direction of the forces when you do this. You should:

- add forces that are acting in the same direction
- subtract forces that are acting in opposite directions.

You need to state the direction as well as the size when you calculate a resultant force.

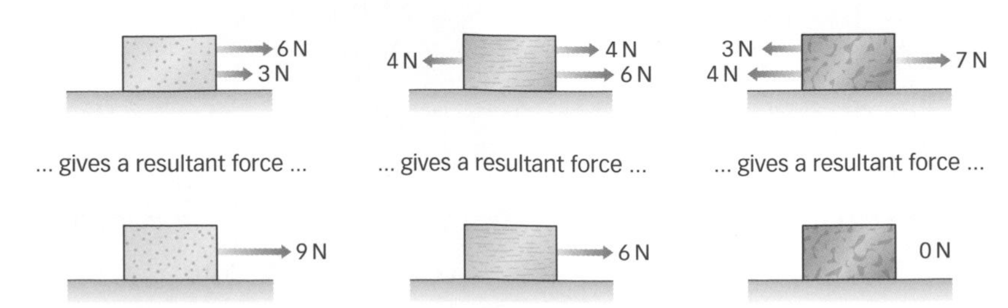

▲ Three forces are replaced by a single resultant force.

Motion

The resultant force on an object determines its motion.

- If the resultant force is zero the motion does not change. The object carries on moving with the same speed and direction (velocity).
- If the resultant force is in the direction of motion, the object speeds up (accelerates).
- If the resultant force is opposite to the direction of motion, the object slows down.

Key words

force, pair of forces, resultant force, acceleration, mass

Acceleration

The **acceleration** of an object depends on its **mass** and resultant force. They are linked by this equation:

$$F = m \times a \text{ or } a = \frac{F}{m}$$

F is the resultant force in newtons (N)
m is the mass in kilograms (kg)
a is the acceleration in metres per second squared (m/s^2)

A car of mass 1500 kg has an acceleration of 6 m/s^2. What is the resultant force on it?

$$F = m \times a = 1500\,\text{kg} \times 6\,\text{m/s}^2 = 9000\,\text{N}$$

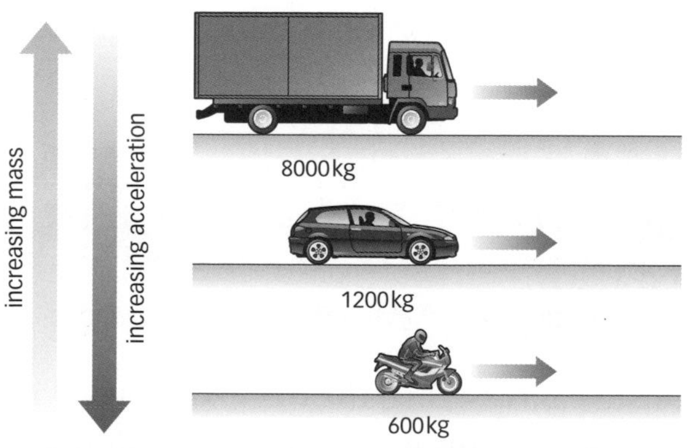

▲ The mass of an object affects its acceleration for the same force.

Exam tip

AQA

Work out the resultant force on an object before deciding how its velocity changes.

Questions

1. Edward sits on a chair. He exerts a downwards force of 600 N on the chair. What force does the chair exert on Edward?
2. A box has three forces acting on it. Forces of 6 N and 12 N push it to the left. A force of 9 N pushes it the right. What is the resultant force on the box?
3. **H** A car has a mass of 900 kg. The engine pushes it forwards with a force of 4000 N. Friction pushes backwards on it with a force of 1000 N. Calculate its acceleration.

4: Forces and braking

Revision objectives

- ✔ know that a constant velocity requires balanced forces
- ✔ understand stopping, braking, and thinking distances for a car
- ✔ understand why braking force increases with speed for a fixed stopping distance
- ✔ state factors that affect stopping distance

Student book references

2.7 Resistive forces
2.8 Stopping distances

Specification key

✔ **P2.1.3 a – d, f**

▲ The horizontal forces on a moving car.

Forces on a car

Two horizontal forces act on a car when it moves along a level road:

- the **driving force** pushes the car forwards
- the **resistive forces** push the car backwards.

The driving force is caused by the engine turning the wheels. There are two different resistive forces. The **friction force** appears whenever objects slide against each other, and **air resistance** appears when objects have to push air out of their way.

Balanced forces

If the driving and resistive forces are the same size, then the resultant force on the car will be zero. This is because the two forces act in opposite directions, so cancel out. When this happens, the forces are **balanced** and the car's velocity remains constant. It doesn't speed up or slow down.

Unbalanced forces

For a car to speed up, the resultant force must be in the forwards direction. This means that the driving force must be greater than the resistive forces. If the driving force is smaller

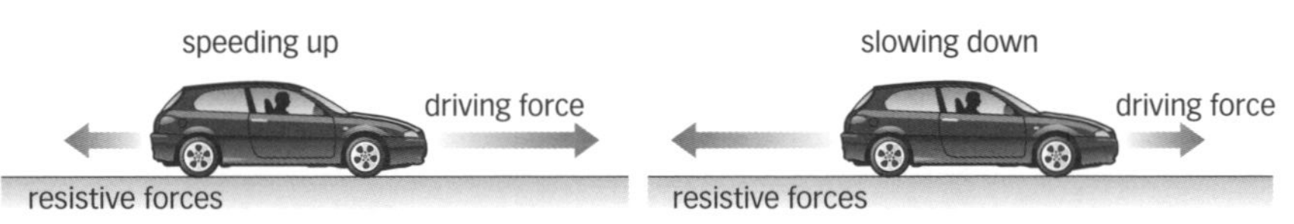

▲ The speed of the car changes if the forces are unbalanced.

than the resistive forces, then the resultant force will be in the backwards direction. This will make the car slow down.

Braking force

When the driver wants to stop their car, they do two things:

- take their foot off the accelerator, making the driving force zero
- put a foot on the brake pedal, adding the **braking force** to the resistive forces.

This applies a large resultant force in the backwards direction to the car. The car slows down until it stops.

▲ Forces on a car when the brakes are applied.

Stopping distance

The **stopping distance** is the total distance that a car moves when it is stopped. Once the driver decides to stop there will be a **reaction time** before they put their foot on the brakes. During this time the car continues moving at a steady speed through the **thinking distance**. Once the brakes have been applied, the car moves at a falling speed through the **braking distance**.

stopping distance = thinking distance + braking distance

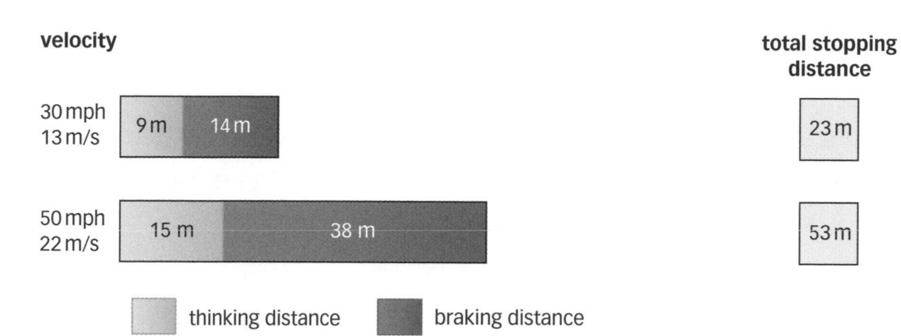

▲ Stopping distance increases with increasing velocity.

Thinking distance

There are ways of increasing the thinking distance:

- driving under the influence of drugs such as alcohol
- distracting the driver with mobile phones or satellite navigation systems.

Any factor that increases the reaction time increases the chance of an accident.

Braking distance

The braking distance increases with increasing speed of the car. This is because the car has to decelerate for longer, so travels further in that time. The braking distance for a given speed can be increased by:

- reducing the friction between road and tyres with ice, snow, or rain on the road
- reducing the braking force by having worn or badly adjusted brakes.

Exam tip

AQA

Don't forget to include a reason as well as a factor when discussing changes of thinking or braking distance.

Key words

driving force, resistive force, friction force, air resistance, balanced, braking force, stopping distance, reaction time, thinking distance, braking distance

Questions

1. The stopping distance for a car at 70 mph is 96 m. If the thinking distance is 21 m, what is the braking distance?
2. Discuss the factors that can increase the stopping distance of a car.
3. **H** A car starts from rest and accelerates to a steady top speed of 80 mph. If the driving force is a steady 800 N, explain how the resistive forces change in this time.

5: Terminal velocity

Revision objectives

- calculate the weight of an object
- describe the motion of an object falling under gravity
- understand how a falling object reaches a terminal velocity
- draw and interpret velocity–time graphs for falling objects

Student book references

2.9 Motion under gravity

2.10 Terminal velocity

Specification key

✓ P2.1.4

Weight

The gravitational force on an object is called its **weight**. You calculate it with this equation:

$$W = m \times g$$

W is the weight of the object in newtons (N)
m is the **mass** of the object in kilograms (kg)
g is the **gravitational field strength** in newtons per kilogram (N/kg)

Since weight is a force it must always have a direction. It always acts towards the centre of the planet. The value of g depends on the size of the planet – on Earth it is about 10 N/kg but on Mars it is only 3.7 N/kg.

Falling down

Gravity can only accelerate objects downwards when they are unsupported. So when the ball is on the table, its weight is cancelled out by an equal and opposite force from the table. As the forces on the ball are balanced its motion cannot change – it just stays there.

Without the table there is only gravity (and a small amount of air resistance) acting on the ball. The resultant force is its weight, so it accelerates downwards.

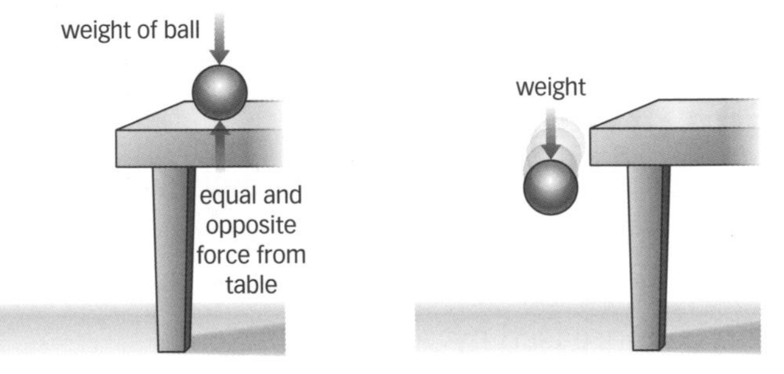

▲ Two equal but opposite forces act on the ball when it is on the table, but there is a resultant force while it falls (small amount of air resistance not shown).

Air resistance

Objects that fall through the air have to push their way through the air. So the air exerts an equal and opposite upwards force on the object: **air resistance**. The size of air resistance can be increased by:

- increasing the speed of the object
- increasing the surface area of the object
- making the shape of the object less streamlined.

So air resistance affects flat, light shapes such as leaves much more than round, heavy ones like balls.

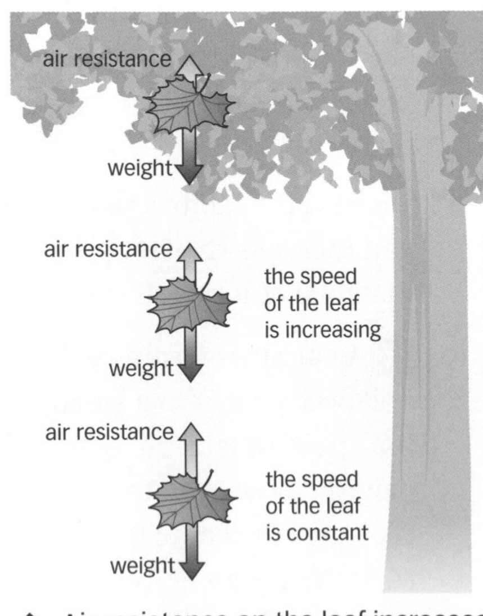

▲ Air resistance on the leaf increases as it speeds up.

Terminal velocity

Air resistance sets a limit to the velocity of a falling skydiver. As the skydiver falls, their weight stays constant but their air resistance increases as they speed up. Since the two forces act in opposite directions, they will eventually become balanced and the skydiver will fall at a steady speed – the **terminal velocity**.

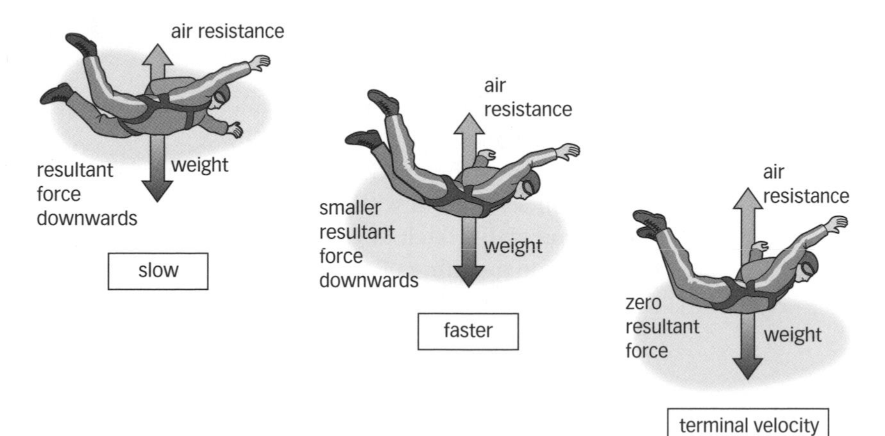

▲ The air resistance on the skydiver increases as they fall through the air.

Parachutes

The terminal velocity of a skydiver can be over 100 mph. This is not a safe landing speed. Usually the skydiver opens a parachute before they hit the ground. The parachute has to push a lot of air out of the way, so it has a much higher air resistance than the skydiver. This extra upwards force on the skydiver slows them down to a new, much smaller, terminal velocity.

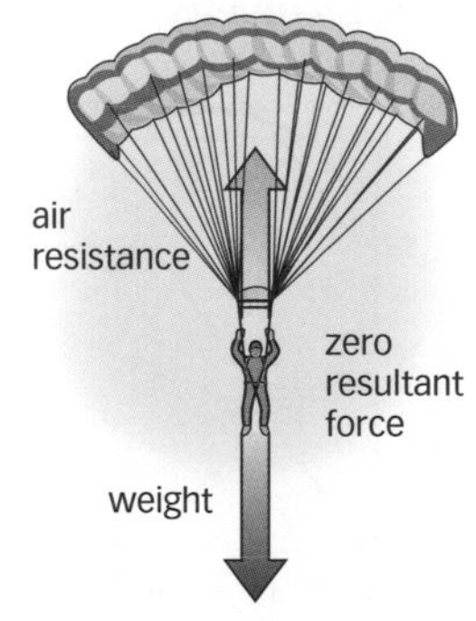

▲ A parachute increases air resistance.

Key words

weight, mass, gravitational field strength, air resistance, terminal velocity

Exam tip

AQA

Remember that *mass* and *weight* are different things. Moving an object from one planet to another does not affect its mass.

Questions

1 Calculate the weight of a 65 kg person on Mars where $g = 3.7$ N/kg.

2 Explain why you are safe when you stand on the edge of a cliff but not when you step over the edge.

3 **H** The velocity–time graph below is for a skydiver who makes a safe landing. Explain the shape and stages of the graph.

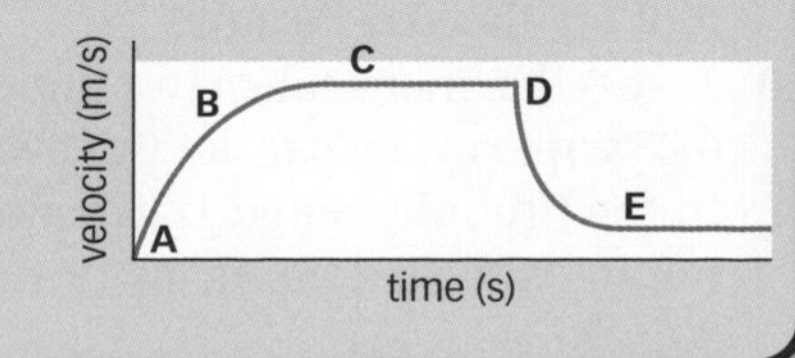

Questions
Force and motion

Working to Grade E

1 State what happens to the motion of a car if:
   a the driving force is greater than the resistive forces
   b the driving force is less than the resistive forces
   c the driving force is equal to the resistive forces.

2 Calculate the resultant force on a car of mass 900 kg that accelerates at $3\,m/s^2$.

3 A plane moves 250 m in just 5 s. How fast is it moving?

4 Use the velocity–time graph of a car, below, to complete the sentences that follow. Use these words (some may be used more than once):
accelerates decelerates steady zero

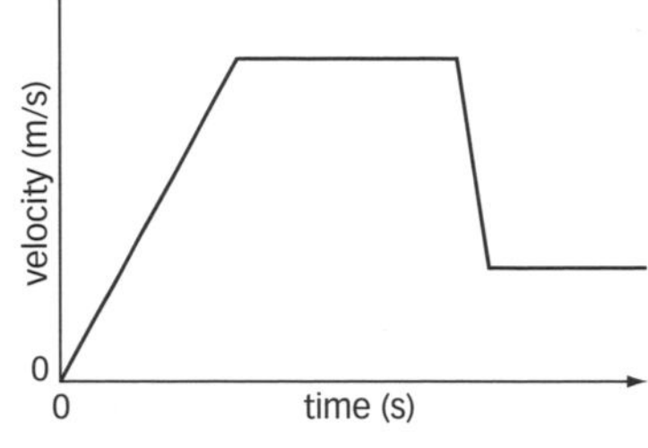

The car starts off with ________ speed. It then ________ to a ________ speed. Then it ________ to a ________ speed.

5 State **two** factors that could increase the stopping distance of a car.

6 Complete the sentences below for a skydiver. Choose from these words: **acceleration air resistance balanced unequal velocity weight**
As the skydiver speeds up, her ________ remains constant but her ________ increases.
When these two forces are ________ she is at her terminal ________ .

Working to Grade C

7 Three forces act on a brick, as shown below. Calculate the size and direction of its acceleration.

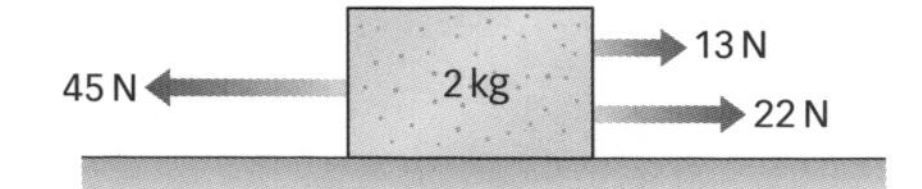

8 Calculate the weight of the brick ($g = 10\,N/kg$) and explain why the brick's vertical acceleration is zero.

9 A car of mass 750 kg moves at steady speed of 10 m/s when the driving force is 250 N. What is the deceleration of the car when the driver takes her foot off the accelerator?

10 A speed camera takes two photos of a passing car 0.25 s apart. The car is 5.6 m away from the camera in the first photo and 12.4 m away in the second photo. Is the car going faster than the speed limit of 30 m/s?

11 Use this distance–time graph of a car to state how its speed is changing with time.

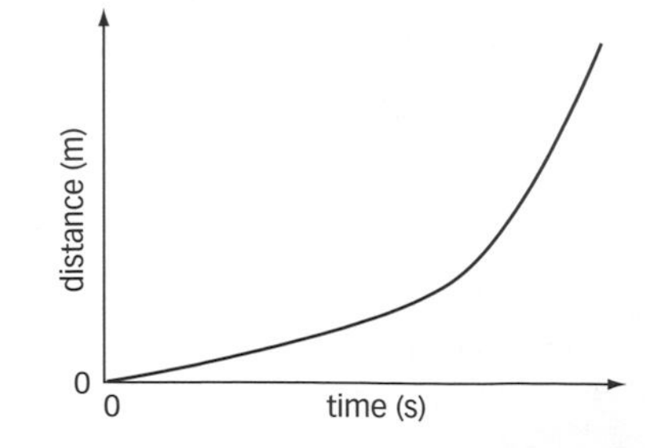

12 Sketch the distance–time and velocity–time graphs for a person who behaves as follows:
- runs away from you at a steady speed
- then stops for a while
- then walks back to you at a steady speed.

13 State the meaning of the following terms: braking distance, stopping distance, and thinking distance.

Working to Grade A*

14 Use this girl's distance–time graph to calculate her velocity when the time is 1 s, 2 s, and 3 s.

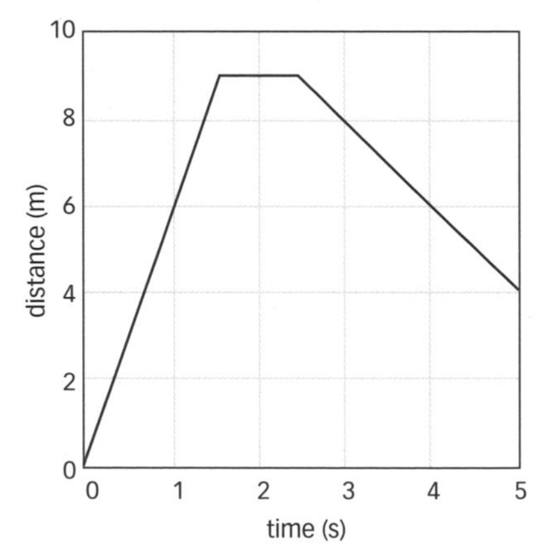

15 Use this car's velocity–time graph to calculate how far it went.

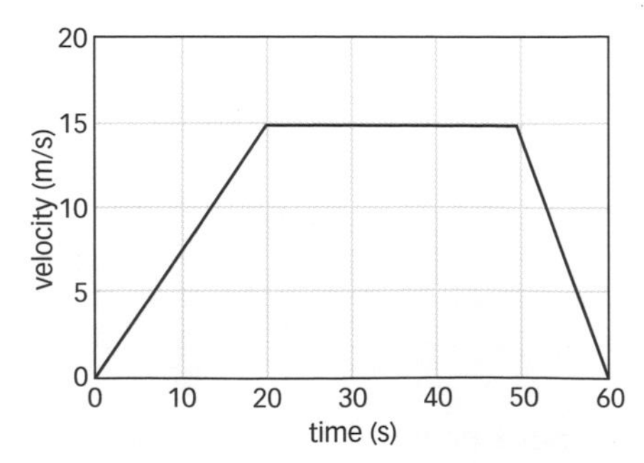

Examination questions
Force and motion

1 The graphs in List A show how the distances of three vehicles change with time.

The statements in List B describe different motions.

Draw one line from each graph in List A to the description of the motion represented by that graph in List B.

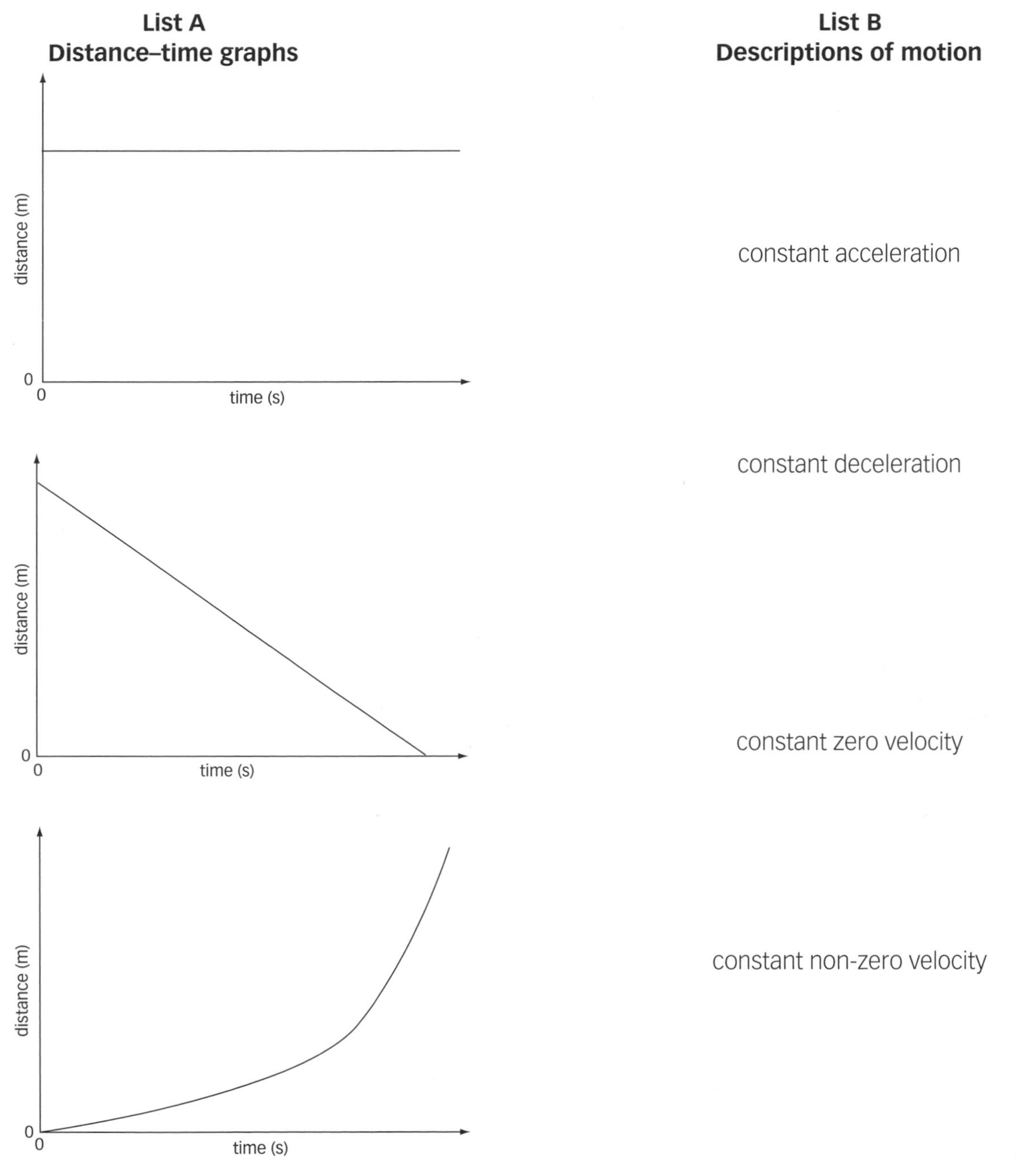

(3 marks)

(Total marks: 3)

2 The diagram shows the four forces acting on a car that is moving along a level road.

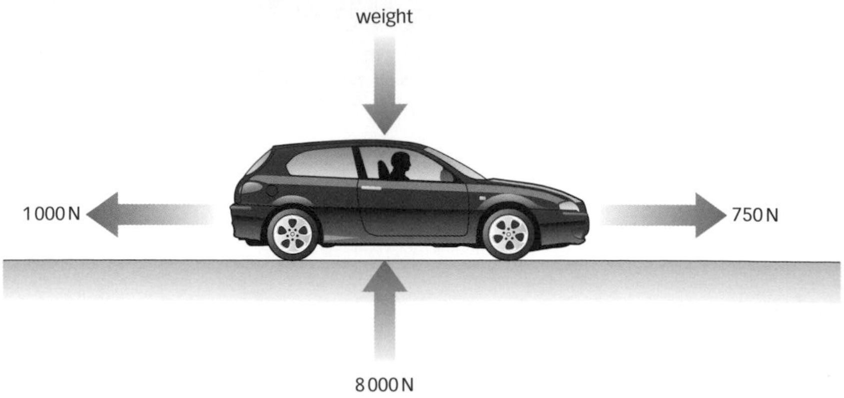

weight

1000 N

750 N

8000 N

a Explain why the weight of the car must be 8000 N.

....

....

(2 marks)

b Calculate the resultant force on the car.

Show clearly how you work out your answer.

....

....

Resultant force = ... N

(2 marks)

H **c** Calculate the acceleration of the car when the forces shown in the diagram act on it.

....

....

....

Acceleration = ... m/s^2

(3 marks)

d Explain how the velocity of the car changes with time when the forces shown in the diagram act on it.

....

....

....

(2 marks)

(Total marks: 9)

3 The table below provides data on thinking and braking distances for a car that does an emergency stop on a dry, level road.

Speed in m/s	Thinking distance in m	Braking distance in m
10	6	8
20	12	32
30	18	72

a Calculate the stopping distance of the car at a speed of 20 m/s.

....

Stopping distance = ... m

(1 mark)

b Use the data in the table above to calculate the reaction time of the driver.

....

....

Reaction time = ... s

(2 marks)

c **i** Explain why the braking distance of the car increases as its speed increases.

....

....

....

....

(3 marks)

ii Explain **one** further factor that affects the braking distance.

....

....

....

(2 marks)

(Total marks: 8)

6: Work and energy

Revision objectives

- understand that work is done when an object is moved against a force
- use the equation *work done = force × distance moved in the direction of the force* ($W = F \times d$)
- understand that work done is equal to energy transferred
- describe the energy transfers when a force is used to change the shape of a spring
- use the equation *force = spring constant × extension* ($F = k \times e$)

Student book references

2.11 Hooke's law

2.12 Work done and energy transferred

Specification key

- **P2.1.5**
- **P2.2.1 a – d**

Work

Whenever an object moves in the direction of an applied force, work is done on it. You calculate the **work done** with this equation:

$$W = F \times d$$

W is the work done on the object in joules (J)
F is the force applied to the object in newtons (N)
d is the distance moved by the object in the direction of the force in metres (m)

So how much work is done on a 150 N weight when it is raised vertically through a distance of 2 m?

$$W = F \times d = 150\,\text{N} \times 2\,\text{m} = 300\,\text{J}$$

Energy transfer

Every time work is done, some energy is transferred. So when the 150 N weight is raised by 2 m, 300 J of energy is transferred from the weightlifter to the weight. 300 J of chemical energy in the weightlifter's muscles transfers to 300 J of **gravitational potential energy** in the weight.

work done = energy transferred

When the weight is lowered back to the ground, it transfers 300 J of gravitational potential energy back to the weightlifter as **heat** in his muscles.

Work against friction

Moving objects have **kinetic energy**. Whenever moving objects slide past each other, there is a frictional force opposing the motion. So work is done whenever friction is present. This work transfers kinetic energy of the moving objects to heat.

Elastic potential energy

When a force squashes or stretches an object, work is done on the object. That work transfers **elastic potential energy** to the object. When the force is removed, that energy is transferred, often as kinetic energy.

Hooke's law

Hooke's law says that the extension of a spring is directly proportional to its **extension**. This is represented in this equation.

$$F = k \times e$$

F is the force extending the spring in newtons (N)
k is the **spring constant** of the spring in newtons per metre (N/m)
e is the extension of the spring in metres (m)

Bungee rope has a spring constant of 250 N/m. So how much force is needed to increase the length of a bungee rope from 25 m to 30 m?

$$F = k \times e = 250\,\text{N/m} \times (30\,\text{m} - 25\,\text{m}) = 1\,250\,\text{N}$$

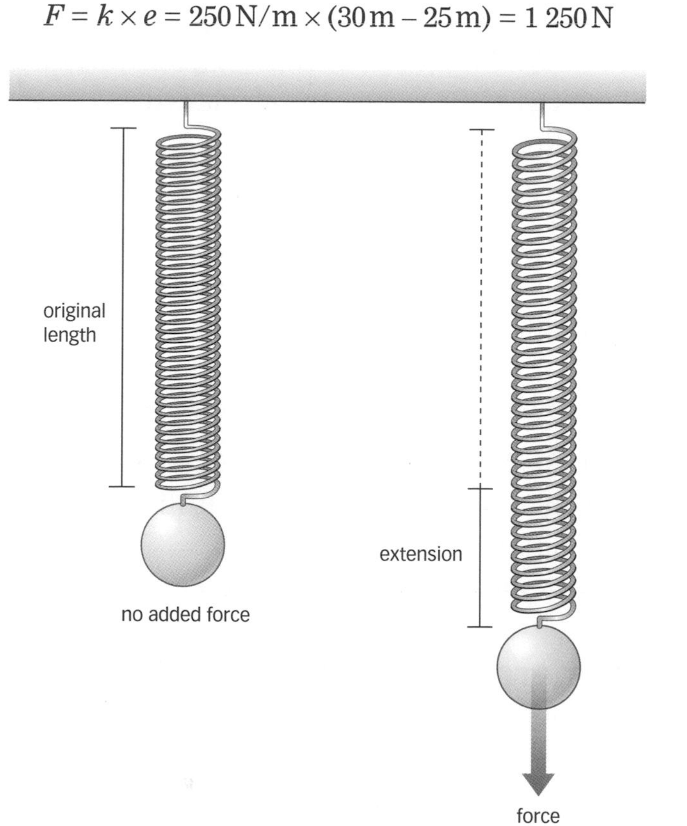

▲ The extension of a spring is its change of length due to the applied force.

Key words

work done, gravitational potential energy, heat, kinetic energy, elastic potential energy, Hooke's law, extension, spring constant

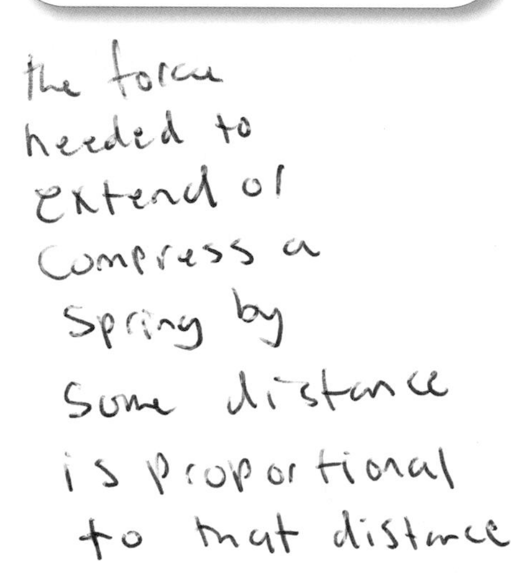

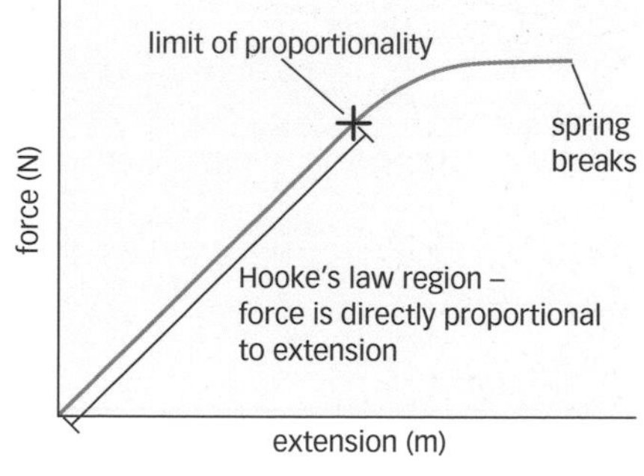

▲ Hooke's law can only be used reliably if the spring isn't stretched too much.

Questions

1. Jim weighs 800 N. He uses the stairs to climb a vertical height of 6 m. How much gravitational potential energy does he gain?
2. How much kinetic energy is transferred to heat when a car brakes in a distance of 40 m by a braking force of 6 000 N?
3. **H** Calculate the spring constant of the elastic band whose length is increased from 10 cm to 25 cm by an applied force of 50 N.

Exam tip AQA

The force and distance must be in the same direction when you calculate work done. Don't confuse the extension of a spring with its length.

7: Energy transfer and power

Revision objectives

- understand what affects the gravitational potential energy of an object
- understand what affects the kinetic energy of an object
- use the equations $E_p = m \times g \times h$ and $E_K = ½ \times m \times v^2$
- understand that power is the rate of doing work
- use the equation $P = \frac{E}{t}$

Student book references

2.13 Power
2.14 Gravitational potential energy and kinetic energy

Specification key

✔ P2.2.1 e – g

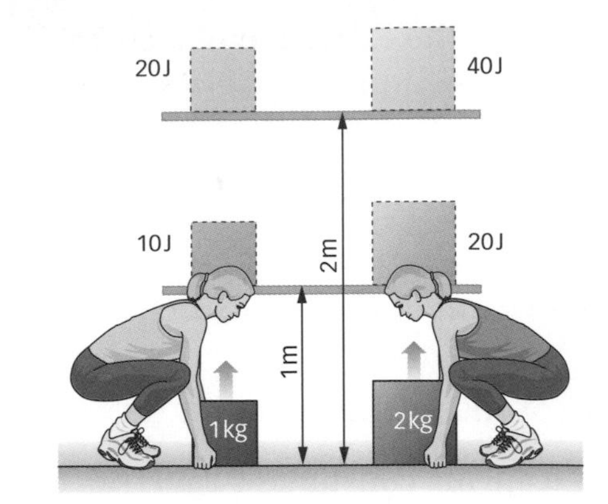

▲ Changes of gravitational potential energy are affected by mass and height.

Gravitational potential energy

Whenever an object is moved up it gains **gravitational potential energy** (GPE). You can calculate the change of GPE with this equation:

$$E_p = m \times g \times h$$

E_p is the change of GPE of the object in joules (J)
m is the **mass** of the object in kilograms (kg)
g is the gravitational field strength in newtons per kilogram (N/kg)
h is the vertical change of height in metres (m)

So how much GPE does a boy of mass 50 kg lose when he drops a vertical height of 0.5 m when he falls off his chair?
(g = 10 N/kg on Earth)

$$E_p = m \times g \times h = 50\,\text{kg} \times 10\,\text{N/kg} \times 0.5\,\text{m} = 250\,\text{J}$$

Kinetic energy

Moving objects have **kinetic energy** (KE). You can calculate kinetic energy with this equation:

$$E_K = ½ \times m \times v^2$$

E_K is the kinetic energy of an object in joules (J)
m is the mass of the object in kilograms (kg)
v is the **velocity** of the object in metres per second (m/s)

So what is the kinetic energy of a 50 kg cheetah moving at 30 m/s?

$$E_K = ½ \times m \times v^2 = ½ \times 50\,\text{kg} \times (30\,\text{m/s})^2 = 0.5 \times 50 \times 900 = 22\,500\,\text{J}$$

Power

Some energy transfers are much quicker than others. The rate at which energy is transferred is known as **power**. You can calculate it with this equation:

$$P = \frac{E}{t}$$

P is the power in watts (W)
E is the transferred energy in joules (J)
t is the time taken for the energy transfer in seconds (s)

So what is the power of a car of mass 900 kg that can accelerate from 0 m/s to 32 m/s in 8 s?

Start off by calculating the energy transferred to kinetic energy:

$$E_K = ½ \times m \times v^2 = ½ \times 900\,\text{kg} \times (32\,\text{m/s})^2 = 0.5 \times 900 \times 1024 = 460\,800\,\text{J}$$

Then calculate the power:

$$P = \frac{E}{t} = \frac{460\,800\,\text{J}}{8\,\text{s}} = 57\,600\,\text{W}$$

Key words

gravitational potential energy, mass, kinetic energy, velocity, power

H Transfer time

The power of a device allows you to work out how long you will need to use it. Suppose you want to use a microwave oven to reheat a mug of coffee. The microwave will need to transfer 114 000 J of energy as heat to the mug. How long will it take if the microwave's power is 950 W?

$$P = \frac{E}{t} \text{ so } P \times t = E \text{ and } t = \frac{E}{P}$$

$$t = \frac{114\,000\,\text{J}}{950\,\text{W}} = 120\,\text{s}$$

PRO*line*
MOD.:ST44
2450MHz
230V ~ 50Hz
MICROWAVE INPUT POWER : 1550 W
MICROWAVE ENERGY OUTPUT : 950W
SERIAL NO. 81000138
MADE IN KOREA
CE
WARNING – HIGH VOLTAGE

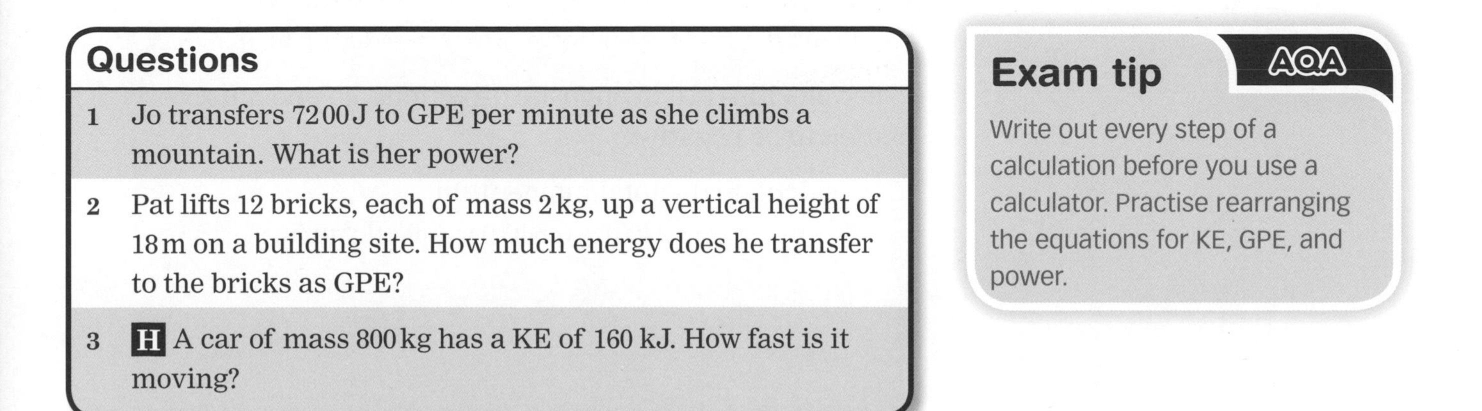

Questions

1 Jo transfers 7200 J to GPE per minute as she climbs a mountain. What is her power?

2 Pat lifts 12 bricks, each of mass 2 kg, up a vertical height of 18 m on a building site. How much energy does he transfer to the bricks as GPE?

3 **H** A car of mass 800 kg has a KE of 160 kJ. How fast is it moving?

Exam tip AQA

Write out every step of a calculation before you use a calculator. Practise rearranging the equations for KE, GPE, and power.

8: Momentum and car safety

Revision objectives

- use the equation $p = m \times v$ to calculate momentum
- apply the law of conservation of momentum
- describe the energy transfers for a braking car
- explain the benefits of air bags and crumple zones for car safety

Student book references

2.8 Stopping distances
2.12 Work done and energy transferred
2.15 Momentum

Specification key

✔ P2.1.3 e ✔ P2.2.1 c
✔ P2.2.2

Momentum

The **momentum** of an object allows you to predict its change of motion when it collides with another object. You calculate its size with this equation:

$$p = m \times v$$

p is the momentum of the object in kilogram metres per second (kg m/s)
m is the mass of the object in kilograms (kg)
v is the velocity in metres per second (m/s)

The momentum has the same direction as the velocity.

Conservation of momentum

The total momentum of a set of interacting objects never changes, providing that the objects only interact with each other. This principle of **momentum conservation** allows you to calculate the effect of collisions or explosions.

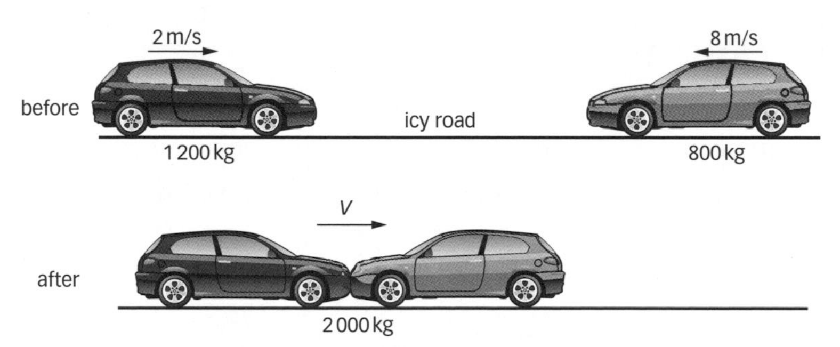

▲ The situation before and after the collision of two cars.

For example, suppose that two cars skid together and have a head-on collision on an icy road. The conservation of momentum allows you to calculate the final velocity V of the cars after they have collided and joined together.

Start off by calculating the total momentum of each car before the collision.

left car: $p = m \times v = 1200\,\text{kg} \times +2\,\text{m/s} = +2400\,\text{kg m/s}$
right car: $p = m \times v = 800\,\text{kg} \times -8\,\text{m/s} = -6400\,\text{kg m/s}$

(The velocities are in opposite directions, so one must be chosen to be negative.)

Then calculate the total momentum:

$$(+2400) + (-6400) = -4000\,\text{kg m/s}$$

After the collision the two cars are locked together and behave like a single object of mass 1200 + 800 = 2000 kg. Provided there is no friction with the icy road, the momentum of this object must be –4000 kg m/s.

$$p = m \times v \text{ so } \frac{p}{m} = v$$

$$V = \frac{-4000\,\text{kg m/s}}{2000\,\text{kg}} = -2\,\text{m/s}$$

So the velocity of the cars after the collision is 2 m/s to the left.

Car safety

When a car has a head-on collision, the people inside have to stop moving. Something has to exert a backwards force on them to reduce their velocity to zero. Normally a **seatbelt** applies this decelerating force. The **crumple zone** at the front of the car increases the time for its velocity to drop to zero. This increase in time reduces the deceleration of the people, and reduces the force needed from the seatbelt.

The time for a collision can be increased in a number of ways:

- allowing the seatbelt to stretch during the accident
- using an **airbag** to slow down heads gradually
- leaving lots of room for the bodywork to collapse.

Energy transfers in braking

When a car is stopped, friction in the brakes transfers the kinetic energy of the car to heat. This is a waste of energy as the heat eventually spreads into the air and cannot be used for anything else. Some cars use **regenerative braking** instead. This slows down the car by transferring the kinetic energy of the car to electricity that charges up a battery. The battery can then transfer this energy back to kinetic energy when the car speeds up again.

Key words

momentum, momentum conservation, seatbelt, crumple zone, airbag, regenerative braking

Exam tip AQA

Remember to take account of directions when calculating total momentum.

Questions

1 Sue has a mass of 52 kg. What is her momentum when she runs at a speed of 7 m/s?

2 Explain how the crumple zone of a car increases the safety of the driver when a car crashes.

3 **H** A gun of mass 2 kg fires a bullet of mass 0.02 kg with a speed of 300 m/s. Calculate the recoil velocity of the gun.

Questions
Work and energy

Working to Grade E

1 A spring has a spring constant of 60 N/m. Calculate the force required to increase its length by 0.25 m.

2 Complete this table with the correct units for each quantity.

Quantity	Unit
kinetic energy	
momentum	
power	
work	

3 Complete the sentences for a braking car.

When the brakes are applied on a level road, the ________ energy of the car transfers to ________ in the brakes. If the car is going uphill at the time, some of the ________ energy transfers to ________ potential energy.

4 A crane can do 100 000 J of work in 20 s. What is its power?

5 A plane of mass 2 000 kg falls a vertical distance of 50 m. What is its change of GPE? (g = 10 N/kg)

6 State **three** features of cars that increase the safety of the driver in a crash.

7 Calculate the momentum of 0.5 kg ball when it has a speed of 20 m/s.

Working to Grade C

8 Explain why a seatbelt increases the safety of the driver when a car crashes.

9 A crane lifts a load of mass 250 kg through a vertical height of 30 m in a time of 15 s.

Calculate the power of the crane.

10 A car of mass 1100 kg accelerates from 0 m/s to 30 m/s in a time of 12 s. Calculate the power of the car.

11 State the law of momentum conservation. Include the conditions under which it applies.

12 State Hooke's law. Include the conditions under which it applies.

13 A spring's length increases from 0.26 m to 0.30 m when an extra force of 32 N is applied. Calculate the spring constant of the spring.

14 Which of these quantities have both a size and a direction?

acceleration force kinetic energy
momentum power velocity

15 Explain why the use of regenerative braking increases the fuel efficiency of a car.

Working to Grade A*

16 A trolley of mass 2 kg moves towards a stationary trolley of mass 3 kg with a velocity of 1.5 m/s. The trolleys collide and stick together. Calculate the velocity of the joined trolleys after the collision, stating any assumptions you have to make.

17 A ball of mass 1.6 kg is released from rest and falls down a deep well through a vertical height of 20 m. Calculate the speed of the ball just before it hits the water. State any assumptions you have to make (g = 10 N/kg).

Examination questions
Work and energy

1 This diagram shows a car moving up a ramp at a steady speed.

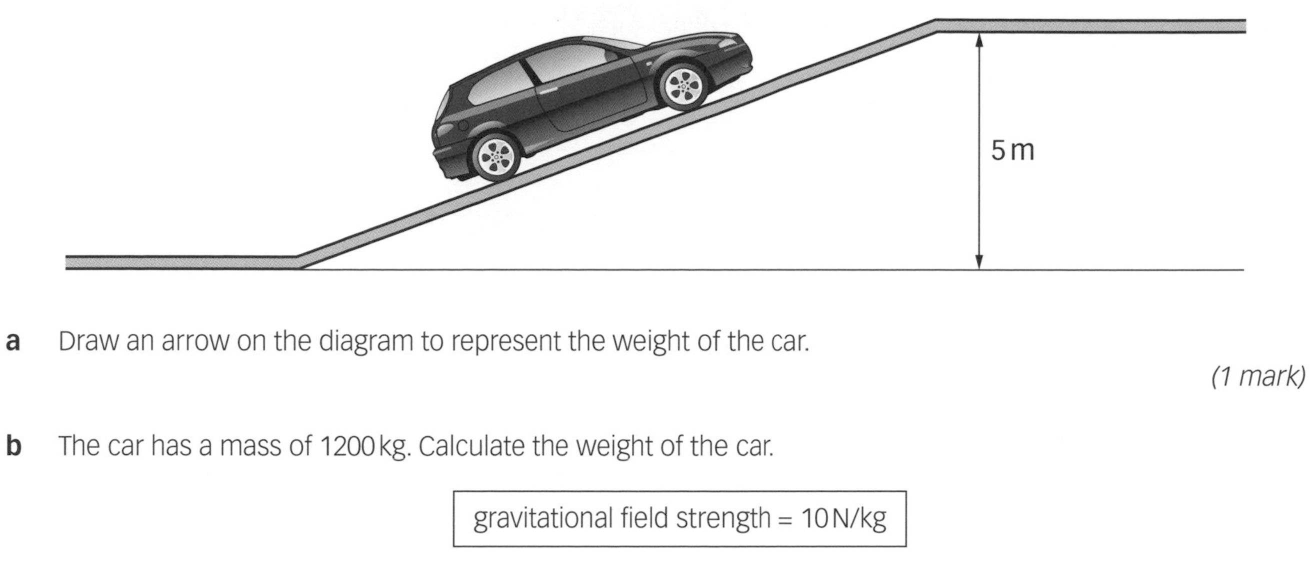

a Draw an arrow on the diagram to represent the weight of the car.

(1 mark)

b The car has a mass of 1200 kg. Calculate the weight of the car.

gravitational field strength = 10 N/kg

Write down the equation you use, and then show clearly how you work out your answer.

....

....

Weight = ... N
(2 marks)

c The ramp raises the car up by 5 m.

Calculate the work done by the car as it moves from the bottom of the ramp to the top.

Write down the equation you use, and then show clearly how you work out your answer.

....

....

Work done = ... J
(2 marks)

d Which **one** of these quantities increases as the car moves up the ramp at a steady speed? Circle your answer.

GPE	KE	momentum	weight

(1 mark)
(Total marks: 6)

2 A car with a mass of 900 kg travels along a level road.

It accelerates from rest to a speed of 15 m/s in a time of 5 s.

a Calculate the kinetic energy of the car at its final speed.

Write down the equation you use, and then show clearly how you work out your answer.

....

....

kinetic energy = ... J

(2 marks)

b Calculate the power developed by the car in the time of 5 s.

Write down the equation you use, and then show clearly how you work out your answer.

....

....

power = ... W

(2 marks)

c *In this question you will be assessed on using good English, organising information clearly, and using specialist terms where appropriate.*

The car crashes into a solid wall.

Explain how the crumple zone of the car increases the safety of the driver in the crash.

....

....

....

....

....

....

....

....

....

....

....

....

(6 marks)

(Total marks: 10)

3 A marksman shoots a bullet into a block of wood.

400 m/s

bullet

block

a The bullet has a mass of 0.025 kg. Calculate the momentum of the bullet.

Write down the equation you use, and then show clearly how you work out your answer.

....

....

momentum = ... kg m/s

(2 marks)

b The bullet enters the block and joins onto it. The block is on a friction-free mounting.

i Which one of these quantities remains the same as the bullet enters the block?

Tick (✓) **one** box.

Total momentum ☐

Total heat energy ☐

Total kinetic energy ☐

(1 mark)

ii The block has a mass of 2.0 kg.

Calculate its velocity once the bullet has stuck into it.

Write down the equation you use, and then show clearly how you work out your answer.

....

....

velocity = ...m/s

(2 marks)

(Total marks: 5)

Designing an investigation and making measurements

In this module there are several opportunities to design investigations and make measurements. These include the acceleration of objects by forces and the behaviour of springs.

As well as demonstrating your investigative skills practically, you are likely to be asked to comment on investigations done by others. The example below offers guidance in this skill area. It also gives you the chance to practise using your skills to answer the sorts of questions that may well come up in exams.

Comparing the skidding distance of different cylinders

Skill – Understanding the experiment

Bill tested the hypothesis that the kinetic energy of a cylinder is proportional to the distance it skids to a halt.

He placed cylinders with their flat side on a level table and propelled them through a light gate.

The light gate measured the speed of each cylinder.

Each cylinder was allowed to skid to a halt after it had passed through the light gate.

Bill measured the distance between the light gate and the centre of the halted cylinder.

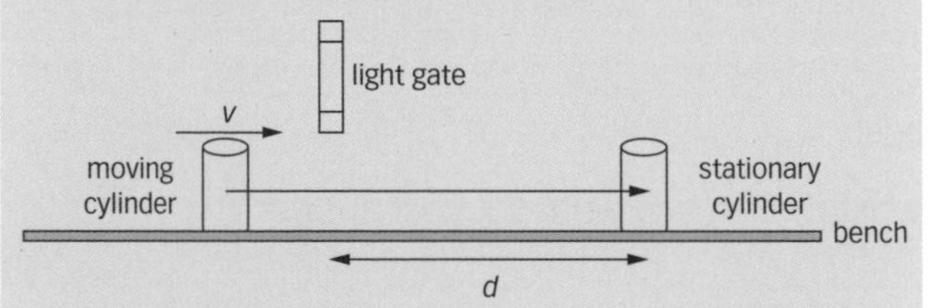

Bill's results are in the table below.

Mass of cylinder (kg)	Initial speed, V (m/s)	Skid distance, d (m)
0.2	3.5	1.23
0.4	2.2	0.56
0.2	1.5	0.27
0.3	3.3	0.98
0.1	2.5	0.76

1 Identify the independent and dependent variables.

In an investigation:

- the independent variables are the ones that are changed by the scientist
- the dependent variable is the one that is measured for each change of the independent variables.

2 Suggest an important control variable for the investigation.

A control variable is one that the investigator thinks might affect the outcome, so they try to keep this variable the same all the way through.

Skill – Using data to draw conclusions

3 What does Bill need to calculate from his results?

Bill has two independent variables that need to be combined to find one of the variables mentioned in his hypothesis.

4 What is the best graph for Bill to draw to see if his hypothesis is correct?

If one quantity is proportional to another, then a scatter graph for the two quantities should be a straight line through the origin.

5 What should Bill do about results that don't fit the same pattern as the others?

Unless there is a very good reason for ignoring data that is well outside the pattern, it should not be ignored. The best thing is to repeat the experiment again to double check the data.

Skill – Societal aspects of scientific evidence

6 Could Bill apply his results to a real situation such as skidding cars?

Bill would need to make estimations from his results to take account of the faster speeds and greater mass of a real car. He might have greater confidence in this if his cylinders and the table had been made from comparable materials.

AQA Upgrade

Answering an extended writing question

QUESTION

In this question you will be assessed on using good English, organising information clearly, and using specialist terms where appropriate.

1 Explain the factors that determine the distance needed to stop a car in an emergency. *(6 marks)*

G–E

When the driver notices the emergency they have to think about putting there foot on the brake before actually doing it, this is the thinking time and the car keeps moving in this time. Then the brakes take over and the speed of the car quickly drops to nothing, this is the braking distance which gets bigger the faster the car is going. It takes longer for the car to stop if the ground is wet, or the driver has been drinking or is using there mobile or is on drugs or too old.

Examiner: This is a poor-quality answer, typical of an E-grade candidate. It is worth 2 marks. There is very little use of technical terms and very little explanation is provided for the statements of fact. It is unclear which distance is affected by which factor. Some spelling is incorrect and sentence structure is often poor.

D–C

There are two stages in stopping a car. The braking distance is how far the car goes while it is actually slowing down. This depends on the speed of the car and how much friction there is between the tyres and the road. The thinking distance is how far the car goes before it starts to slow down. This depends on the reaction time of the driver, which is affected by tiredness and drugs.

Examiner: This is a medium-quality answer, typical of a D-grade candidate. It earns 3 marks.

Although correct technical terms have been used, and most of the factors have been stated, there is no use of the words increase or decrease to show the effect of these factors. Very little explanation of the effect of these factors is provided. The quality of English is good, with no spelling mistakes.

B–A*

The stopping distance is equal to the thinking distance added to the braking distance. The thinking distance is how far the car moves before it starts to slow down, while the driver is reacting to the emergency but before they have got their foot on the brake. The thinking distance is equal to the speed multiplied by the reaction time, so gets longer if either or both the speed and reaction time are increased. Reaction times can be increased if the driver isn't paying attention, drugged up or drunk. Braking distance depends on the speed of the car, its mass, and on the friction in the brakes and the road surface. Reducing the friction reduces the force decelerating the car, and so reduces the deceleration. The time needed for the brakes to stop the car is its speed divided by deceleration, and the longer the time is, the further the car will go. Poor friction can be the result of rain or poor maintenance of the car. Increasing the mass of a car increases the energy which must be transferred by the brakes, so the brakes must be applied for longer.

Examiner: This is a high-quality answer, typical of an A-grade candidate. It is worth 6 marks.

All parts of the question have been explained in sufficient detail. The conclusion on the effect of mass is incorrect, but the candidate has not been penalised for this common error. The candidate has used correct specialist terms and the spelling, punctuation, and grammar are very good.

Part 1 Course catch-up
Motion and energy

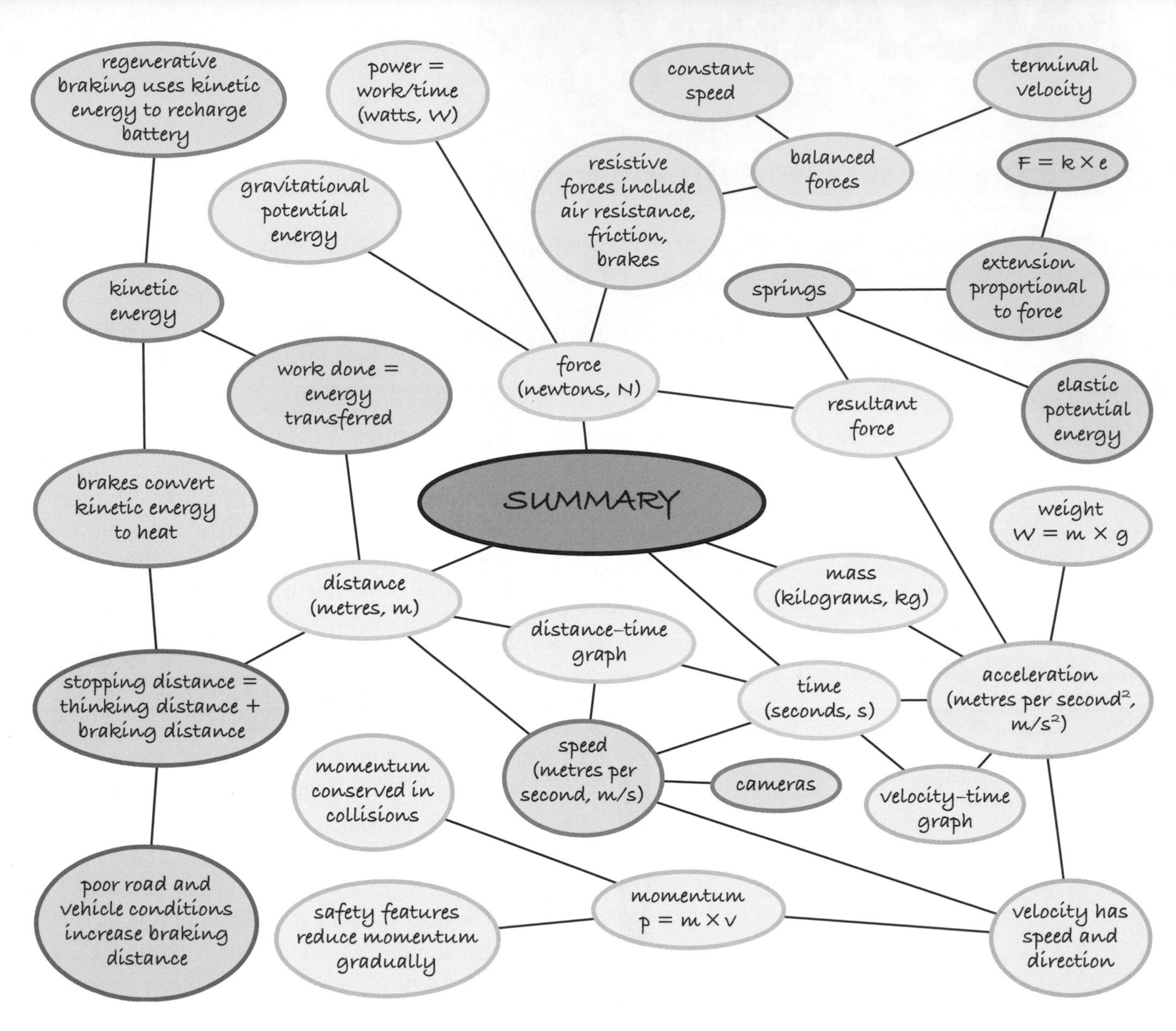

Revision checklist

- Forces are pushes and pulls. Objects exert equal and opposite forces on one another. Resultant force is the sum of all the forces acting on an object.
- Resultant forces change the motion of objects, causing them to accelerate. Resultant force = mass × acceleration.
- Speed is a measure of how fast something is moving. Speed = distance/time. This principle is used in speed cameras.
- Velocity describes direction as well as speed.
- The gradient of a distance–time graph represents speed. Acceleration is a change in speed, and can be positive (speeding up) or negative (slowing down).
- The gradient of a velocity–time graph represents acceleration. The area under a velocity–time graph represents distance travelled.
- A vehicle must overcome the forces of friction and air resistance in order to start moving. When these are balanced by the vehicle's driving force, the vehicle is travelling at a steady speed.
- Stopping distance (thinking distance plus braking distance) is affected by speed, road conditions, vehicle condition, and driver reaction time.
- Weight = mass × gravitational field strength.
- As an object falls under gravity, its velocity increases until its weight is balanced by the force of air resistance. The object then reaches a terminal velocity (steady speed).
- Forces can change the shape of objects. Squashing or stretching a spring gives it elastic potential energy. Hooke's law dictates that the force applied to an object is directly proportional to its extension, up to a certain point.
- Work done is calculated using the equation, work done = force × distance moved in the direction of the force.
- Power = work done/time taken.
- Gravitational potential energy (GPE) is defined by mass, height above the ground, and the Earth's gravitational field. GPE is transferred into kinetic energy (KE) when an object moves.
- A moving object's KE is defined by its mass and its speed.
- The momentum of a moving object is defined by its mass and its velocity.
- Car safety devices increase the time taken for the change in momentum experienced in a crash.

Electron transfer

All materials contain **electrons**. These are small particles that carry negative **charge**. When different materials are rubbed against each other, electrons can be transferred from one to another. The material that gains electrons becomes **negatively** charged. The material that loses electrons becomes **positively** charged.

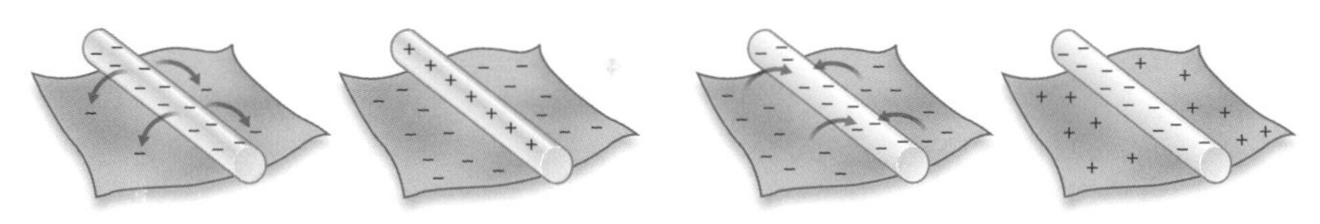

▲ Rubbing two different insulators with a woollen cloth.

The direction of electron transfer depends on the materials used. So when wool is rubbed against polythene, it loses electrons. If wool is rubbed against acetate, it gains electrons instead.

Electrostatic forces

When two electrically charged objects are brought together they exert a force on each other. There are three rules about the forces between charged objects:

- Objects that have the same charge are repelled by each other.
- Objects with different charges are attracted to each other.
- Objects with no charge at all are attracted to charged objects.

Insulators and conductors

Materials can only be charged this way if they are **insulators**. This means that once the electrons have been transferred, they cannot move through the material back to where they started. Electrons can move easily through **conductors**, such as metals, so it is difficult to charge them by rubbing.

Questions

1 What is the difference between a conductor and an insulator?

2 A glass rod is rubbed with a cloth. Electrons transfer from the cloth to the glass. State the charge of the rod and the cloth at the end of this process.

3 **H** A girl rubs a balloon on her hair. This makes her hair stand on end, but also move towards the balloon. Explain these observations.

Revision objectives

- ✔ describe how some materials can be charged by rubbing them
- ✔ explain static electricity with electrons
- ✔ understand that like charges repel and opposite charges attract

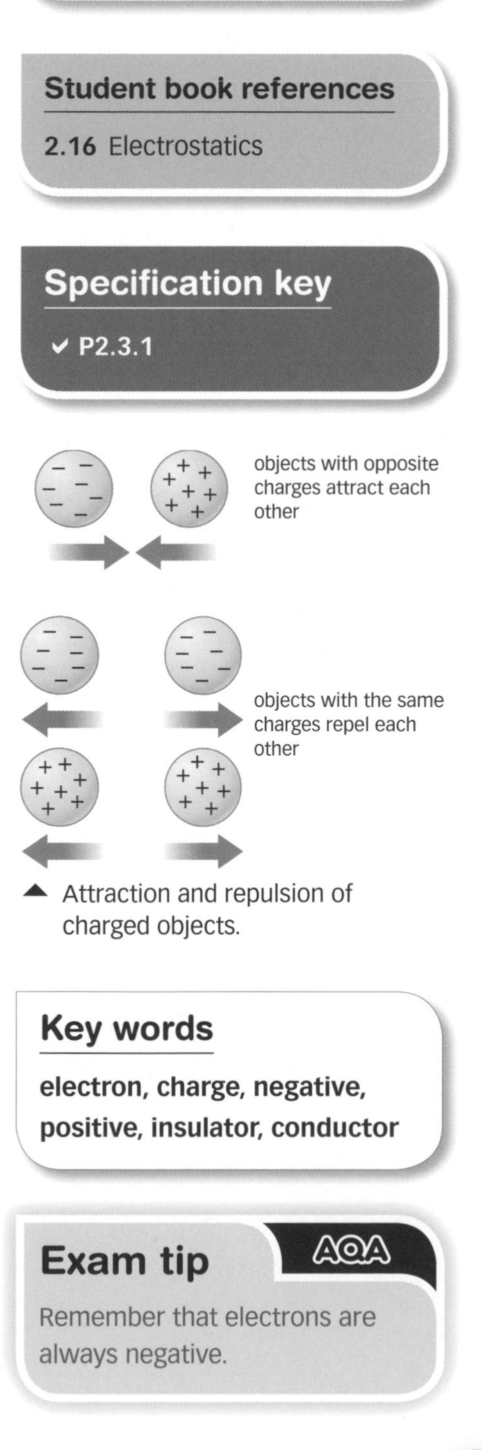

Student book references

2.16 Electrostatics

Specification key

✔ **P2.3.1**

▲ Attraction and repulsion of charged objects.

Key words

electron, charge, negative, positive, insulator, conductor

Exam tip

AQA

Remember that electrons are always negative.

P2

10: Current electricity

Revision objectives

- understand that electric current is the rate of flow of charge
- calculate the size of an electric current
- understand potential difference
- recognise standard circuit symbols
- draw and interpret circuit diagrams

Student book references

2.17 Current and potential difference

2.18 Circuit diagrams

Specification key

- P2.3.2 a – c

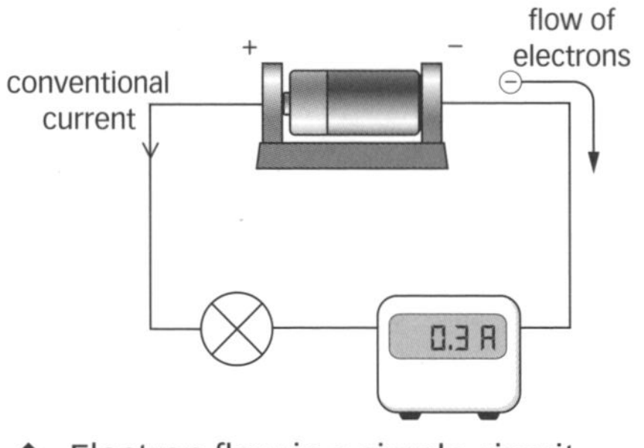

▲ Electron flow in a simple circuit.

Current in circuits

The **cell** in the circuit below pushes electrons around it. The lamp, ammeter, and connecting wires are all conductors, so the electrons can move easily through them. This means that there is a flow of negative charge in the circuit, a **current**. The size of the current in a component tells you how quickly the **charge** is passing through it. You calculate it with this equation:

$$I = \frac{Q}{t}$$

I is the current in amperes (A)
Q is the charge passing through in coulombs (C)
t is the time taken in seconds (s)

So if 480 C passes through the lamp in just 120 s, what is the current in it?

$$I = \frac{Q}{t} = \frac{480\,\text{C}}{120\,\text{s}} = 4\,\text{A}$$

Conventional current

Early scientists guessed that charge in a circuit flowed from the positive terminal of a cell to its negative terminal. This is called the direction of **conventional current**. Nowadays we know that the electrons actually flow the other way, but current is still shown as a flow from positive to negative.

Potential difference

The cell applies a **potential difference** across the circuit. It does this by transferring energy to charge that passes through it. As that charge flows around the circuit it transfers energy to each component that it passes through. You calculate potential difference with this equation:

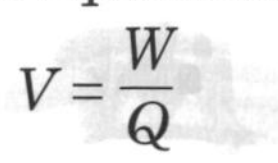

$$V = \frac{W}{Q}$$

V is the potential difference across a component in volts (V)
W is the energy transfer to the component in joules (J)
Q is the charge passing through the component in coulombs (C)

So if 270 J of energy transfer to a mobile phone when 30 C of charge pass through, what is the potential difference across it?

$$V = \frac{W}{Q} = \frac{270\,\text{J}}{30\,\text{C}} = 9\,\text{V}$$

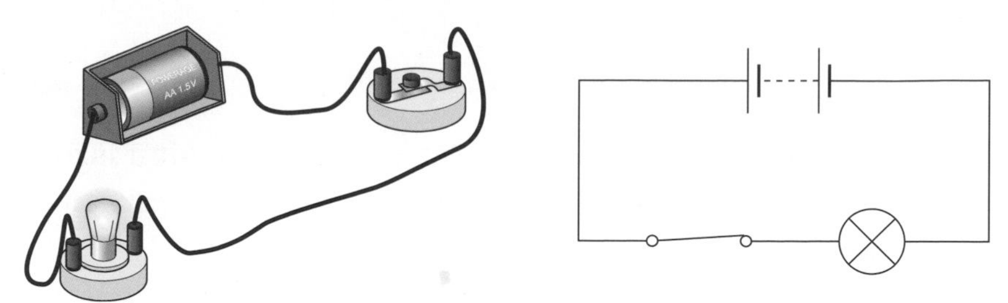

▲ The circuit diagram shows how to assemble the circuit.

Key words

cell, current, charge, conventional current, potential difference, circuit diagram, circuit symbol

Circuit diagrams

Real electrical circuits are usually represented in **circuit diagrams.** Each component is drawn with the correct **circuit symbol**. This can help you to understand how a circuit works.

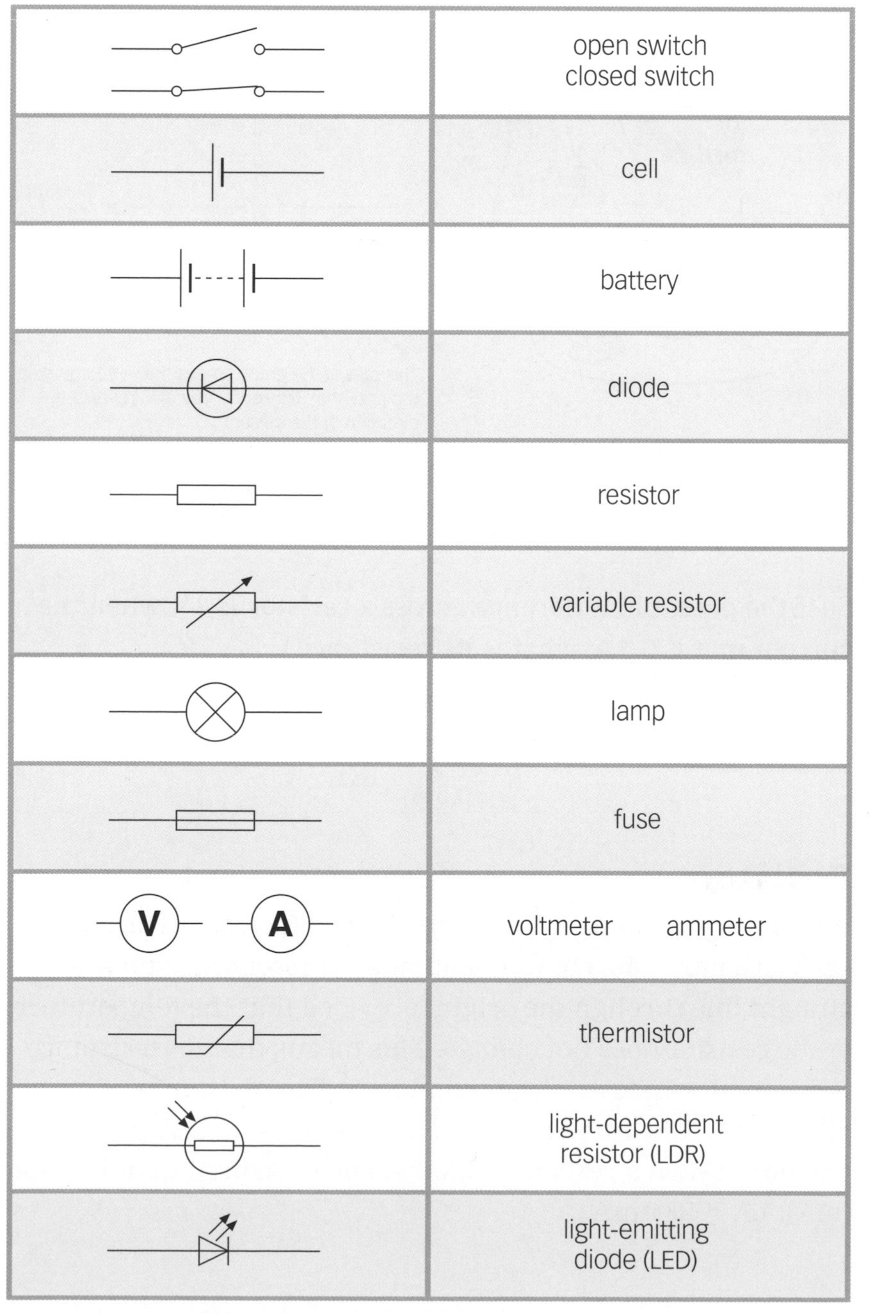

▲ Some circuit components and their symbols.

Exam tip AQA

Make sure that you use the correct symbols when you draw circuit diagrams.

Questions

1 A charge of 27 C flows through a lamp in 9 s. Calculate the current in it.

2 Draw a circuit diagram for a cell connected in series to an ammeter, a resistor, and a diode.

3 **H** The current of 0.5 A in a resistor is switched on for a time of 30 s. If the potential difference across the resistor is 12 V, how much energy is transferred to the resistor in that time?

11: Resistance

Revision objectives

- calculate current, potential difference, and resistance
- draw and interpret current–potential-difference graphs for resistors
- explain how to find the resistance of a component
- draw and interpret current–potential-difference graphs for lamps and diodes
- describe the behaviour of light-dependent resistors and thermistors

Student book references

2.19 Current–potential-difference graphs
2.22 Lamps and light-emitting diodes (LEDs)
2.23 Light-dependent resistors (LDRs) and thermistors

Specification key

✔ P2.3.2 d – f, h, m – q

Resistance

The **resistance** of a component indicates how difficult it is for charge to flow through it. The resistance of a component can be calculated with this equation:

$$V = I \times R$$

V is the potential difference across the component in volts (V)
I is the current in the component in amperes (A)
R is the resistance of the component in ohms (Ω)

To measure the resistance you need to connect an ammeter before the component and connect a voltmeter across both ends of it, as shown in the diagram below.

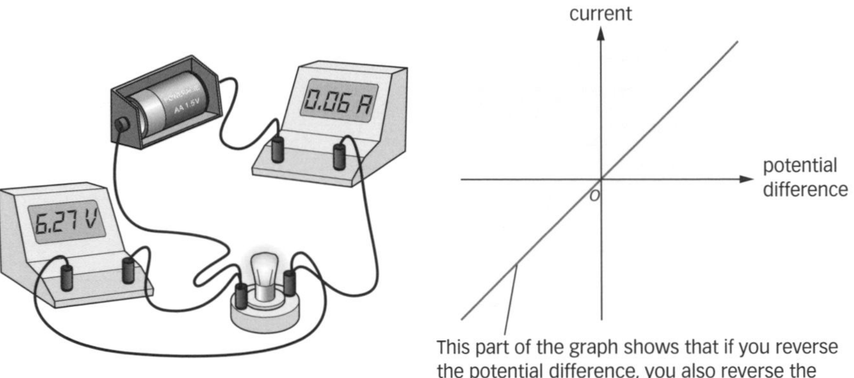

▲ Measuring the current–potential-difference graph for a resistor at constant temperature, illustrating that the current through a resistor is directly proportional to the potential difference across it.

So if the potential difference across a resistor is 3 V when the current in it is 0.5 A, what is its resistance?

$$V = I \times R \text{ so } I \times R = V \text{ and } R = \frac{V}{I}$$

$$R = \frac{3\,\text{V}}{0.5\,\text{A}} = 6\,\Omega$$

Resistors

A **resistor** is usually a thin wire of a conducting material, such as a metal. Its current–potential-difference graph is a straight line through the origin, provided that the temperature of the resistor does not change. This means that its resistance is constant, whatever the current in it. The resistance of a resistor is often quoted in kilo-ohms (kΩ): 1 kΩ = 1000 Ω. The current in such a resistor is likely to be measured in milliamps (mA): 1 A = 1000 mA.

Key words

resistance, resistor, filament lamp, diode, forward, reverse, light-dependent resistor, thermistor

Lamps

A **filament lamp** emits light when there is enough current in it. The charge flows through a thin wire, heating it up enough for it to glow brightly. The increase in temperature increases the resistance of the wire, so its current–potential-difference graph is not a straight line.

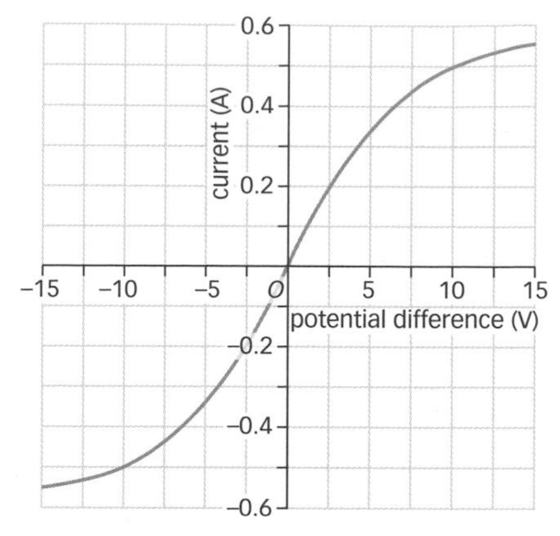

Current–potential-difference graph for a filament lamp.

> **H** The metal in the lamp consists of ions and electrons. When there is a current, the electrons have to flow past the ions, which are fixed in position. As the temperature of the metal increases, the ions vibrate more, getting in the way of the electrons. So the resistance increases as the wire heats up.

Diodes

Charge can only flow one way through a **diode**. The resistance is low when the current is in the **forward** direction, but is very large when the current is in the **reverse** direction.

Some diodes called light-emitting diodes (LEDs) are increasingly being used for lighting. They transfer much less energy as heat compared with filament lamps.

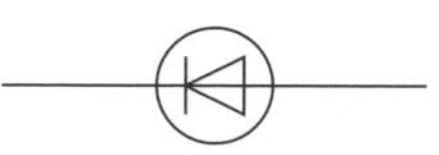

Charge flows easily from right to left through this diode.

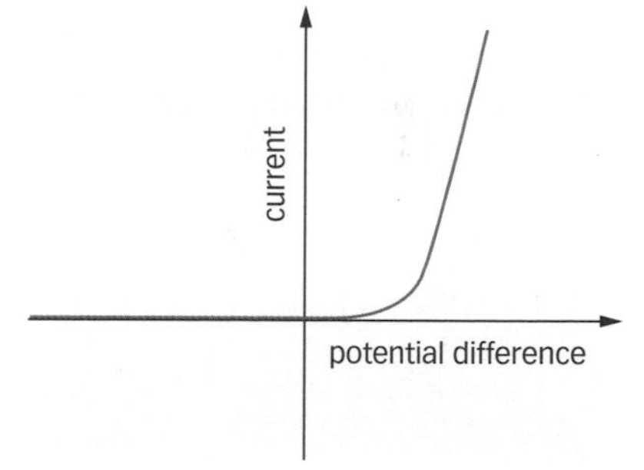

Current–potential-difference graph for a diode.

Sensors

Some resistors can be used as sensors in electrical circuits.

- A **light-dependent resistor** (LDR) has a resistance that decreases with increasing light intensity.
- A **thermistor** has a resistance that decreases as the temperature increases.

Questions

1. What is the potential difference across a 5.6 Ω resistor when it has a current of 0.25 A?
2. Sketch the current–potential-difference graph for a diode and state what it means for the resistance of the diode.
3. **H** Sketch the current–potential-difference graph for a filament lamp. Explain its shape.

Exam tip AQA

Practise changing the subject of the equation $V = I \times R$.

Learn the current–potential-difference graphs for resistors, diodes, and lamps.

12: Circuits in series and parallel

Revision objectives

- recognise series and parallel circuits
- explain how current and potential difference are shared in series circuits
- explain how current and potential difference are shared in parallel circuits
- calculate the resistance of resistors connected in series
- understand the effect of connecting cells in series

Student book references

2.17 Current and potential difference
2.20 Series circuits
2.21 Parallel circuits

Specification key

✓ P2.3.2 h – l

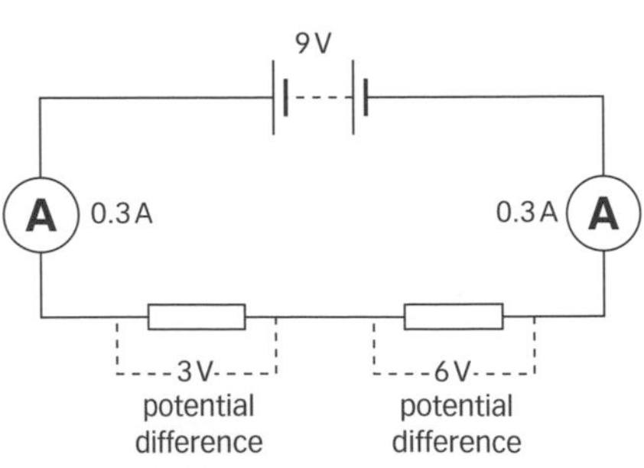

▲ All components have the same current, but only a share of the supply's potential difference.

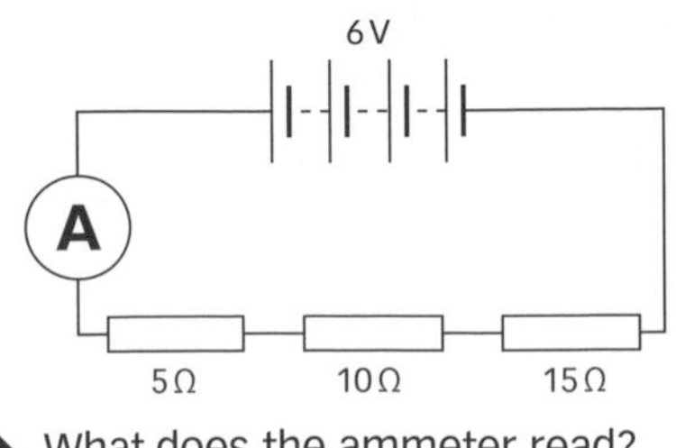

▲ What does the ammeter read?

Series

In a **series** circuit, there is only one way for electrons to get round the circuit. They have to pass through each component in turn.

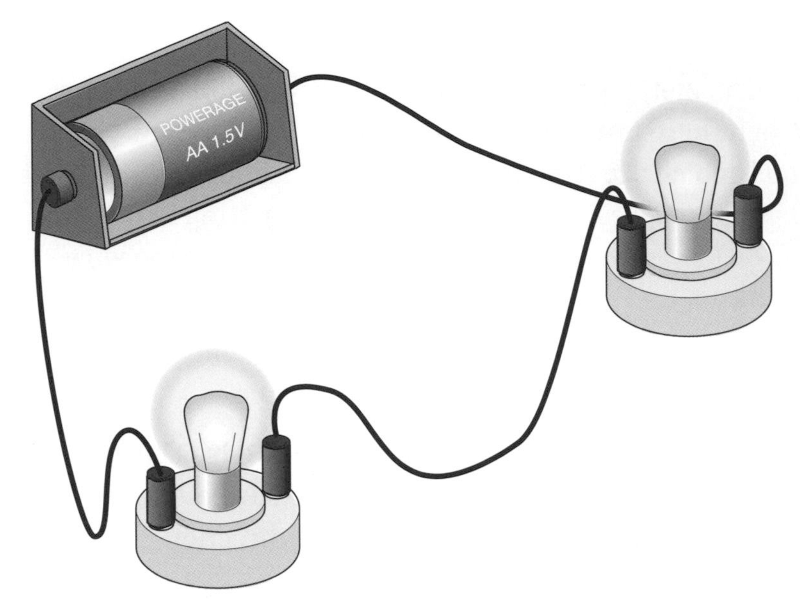

▲ Two lamps in series with a cell.

Current and potential difference

In a series circuit:

- all the components have the same **current**
- the components share the **potential difference** from the supply.

This is because each electron has to pass through each component in turn. So as an electron flows from one terminal of the supply to the other, all of its energy has to be shared amongst the components it passes through.

Resistance

Each component in a series circuit will have a **resistance**. By adding all of their resistances together you find the **total resistance** of the circuit. This allows you to calculate the current in the circuit.

So what is the current in this circuit?

The total resistance = 5 Ω + 10 Ω + 15 Ω = 30 Ω

$$V = I \times R \text{ so } I \times R = V \text{ and } I = \frac{V}{R}$$

$$I = \frac{6\,\text{V}}{30\,\Omega} = 0.2\,\text{A}$$

Parallel

In a **parallel** circuit, there is more than one way for the electrons to get round the circuit.

Current and potential difference

In this circuit, electrons flow through either the resistor or the lamp on their way from one terminal of the battery to the other. They can't flow through both. So although the current leaving the battery is the same as the current entering it, it has to be shared by the components. The lamp has a lower current because it has a higher resistance than the resistor.

Both components have the same potential difference across them. This is because they are directly connected to the supply – there is nothing else for an electron to transfer its energy to.

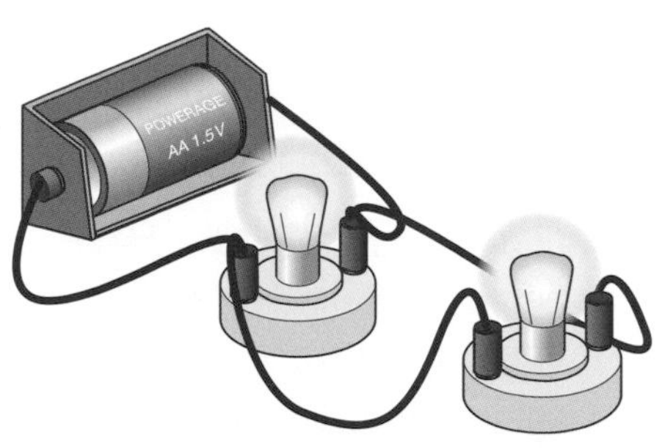

▲ Two lamps in parallel with a cell

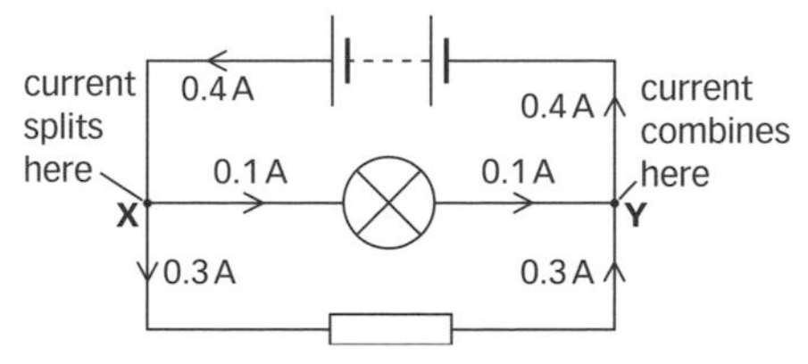

▲ All components have the same potential difference but share the current.

Batteries of cells

When **cells** are stacked in series to make a **battery**, the potential difference across its terminals is the sum of the potential differences across the individual cells. Usually, all of the cells are the same way round. If any cells are the wrong way round, you have to subtract their potential difference from the total. Connecting cells in parallel doesn't increase the potential difference, but does make a battery that lasts longer.

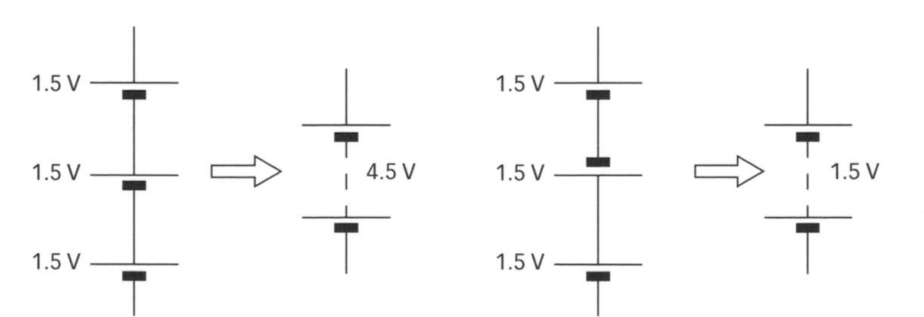

▲ Two batteries of three cells.

Questions

1. A resistor is connected in series with a cell and an ammeter. A voltmeter is connected in parallel with the resistor. Draw the circuit diagram.
2. Resistors of 6 Ω and 3 Ω are connected in series with a 1.5 V cell. By calculating the total resistance, find the current and potential difference for each resistor.
3. **H** Resistors of 6 Ω and 3 Ω are connected in parallel with a 1.5 V cell. By calculating the current in each resistor, find the total resistance of the circuit.

Key words

series, current, potential difference, resistance, total resistance, parallel, cell, battery

Exam tip

AQA

Remember that an ammeter has a low resistance and goes in series with a component, but a voltmeter has a high resistance and goes in parallel with a component.

Revision objectives

- understand the difference between alternating and direct current
- know that mains electricity in the UK is 230 V a.c at 50 Hz
- use an oscilloscope to measure a.c. supply frequency
- understand the structure of a UK three-pin plug
- explain the use of fuses and circuit breakers for electrical safety
- understand why devices are earthed

Student book references

2.24 Direct current and alternating current

2.25 Mains electricity in the home

Specification key

- P2.4.1

Direct and alternating current

If the power supply of a circuit is a battery, then it has **direct current (d.c.)**. The current is always in the same direction, from the positive terminal of the battery to the negative one.

If the power supply of a circuit is a generator, then it has **alternating current (a.c.)**. The current keeps changing direction, first one way, then the other.

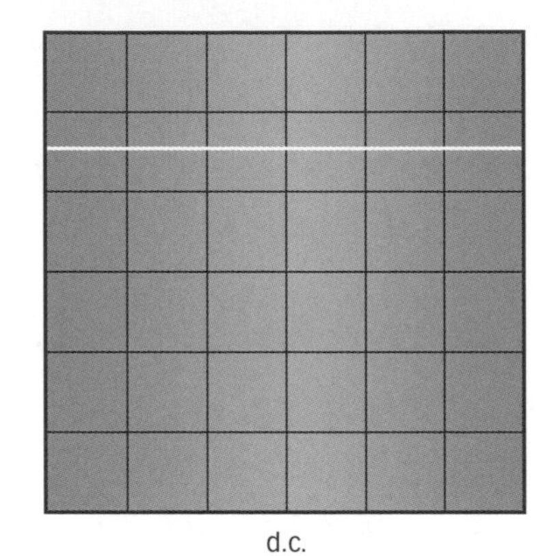
d.c.

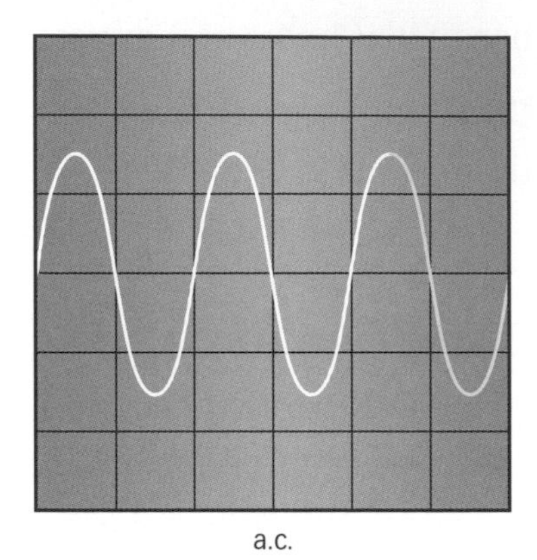
a.c.

▲ Oscilloscope traces for d.c. and a.c. supplies.

Oscilloscope

An **oscilloscope** is an instrument that shows how potential difference changes with time. When the oscilloscope is connected to a d.c. supply, the **trace** on the **screen** is a horizontal line. An a.c. supply gives a wavy line as the potential difference alternates between positive and negative values.

Mains electricity

Mains electricity in the UK is supplied at 230 V a.c. with a **frequency** of 50 hertz or Hz. This means that the potential difference goes through 50 cycles of positive and negative values in each second.

H

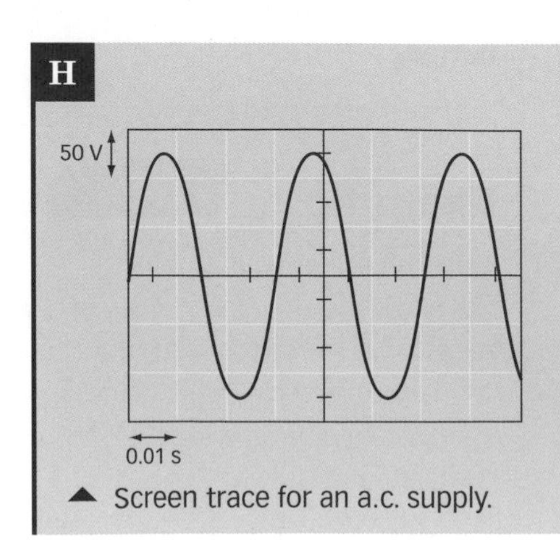

▲ Screen trace for an a.c. supply.

Finding frequency

The trace shows that one complete cycle of change takes up three squares.

The **timebase** of the oscilloscope is set to 0.01 s per square. So the time for one cycle (the **time period**) = 3 × 0.01 = 0.03 s.

$$\text{frequency (Hz)} = \frac{1}{\text{time period (s)}} = \frac{1}{0.03\,\text{s}} = 33\,\text{Hz}$$

The trace also shows that the potential difference alternates between +125 V and –125 V.

UK plugs

UK plugs have three pins that connect with the mains supply.

- The **live** pin carries current at 230 V a.c. from the supply.
- The **neutral** pin allows current to return at a low voltage to the supply.
- The **earth** pin allows current to the ground when something goes wrong.

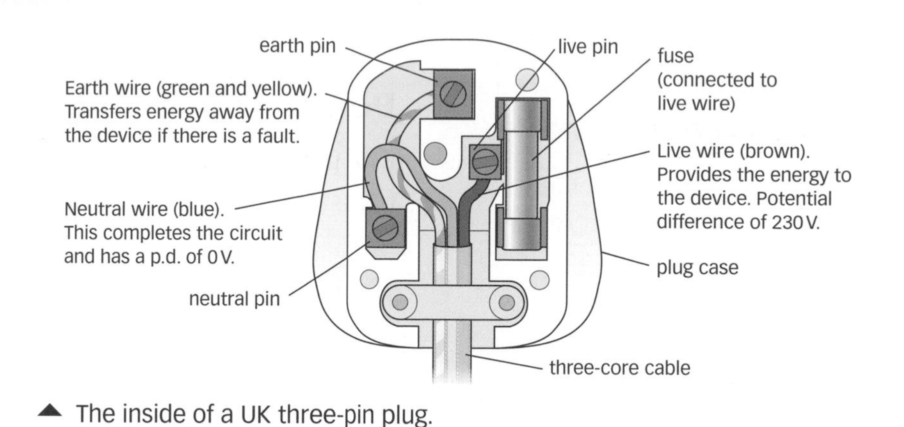

▲ The inside of a UK three-pin plug.

Cable

The cable connecting the plug to the device can have two or three **cores**. These are copper wires surrounded by coloured plastic insulation. The colour of the insulation is linked to the function of the wire inside it.

- The live wire is brown.
- The neutral wire is blue.
- The earth wire is striped yellow and green.

Two-core cables can only be used with devices that are double insulated. This means that if the live wire comes loose it can't connect with the user of the device.

Fuses

Every plug must contain a **fuse**. This is a short length of thin wire in a glass or ceramic tube. It is inserted between the live pin and the live wire. If there is too much current in the live wire, the fuse wire heats up and melts, breaking the circuit. So the fuse is a fast switch that automatically turns off the supply if the current gets dangerously big. A **circuit breaker** is a type of fuse that can be reset by pressing a button.

Earth

Many useful devices have metal casings that are connected to the earth wire. If the live wire accidentally touches the casing, the current can safely escape to earth. Since the earth wire has a low resistance, the current will be large enough to melt the fuse, turning off the supply instantly. A **residual current circuit breaker (RCCB)** does the same job by switching off the supply as soon as the currents in the live and neutral wires are not the same.

Key words

direct current (d.c.), alternating current (a.c.), oscilloscope, trace, screen, frequency, timebase, time period, live, neutral, earth, cores, fuse, circuit breaker, residual current circuit breaker (RCCB)

Exam tip

AQA

Remember that an oscilloscope trace is a potential-difference–time graph. It may look like a wave, but it isn't one.

Questions

1. Describe the structure of a three-core cable and explain the choice of materials for it.
2. What is different between d.c. and a.c. supplies?
3. **H** Explain how the fuse and earth wire in a plug make a device safe to touch.

14: Electrical power

Revision objectives

- use the heating effect of current to explain cable size
- calculate the power of a device
- calculate energy transfer from charge and potential difference

Student book references

2.26 Current, charge, and power

Specification key

✔ P2.4.2

Exam tip

AQA

Always use standard form when you have to calculate with very large or very small numbers. Remember that 1 kW = 1000 W.

Questions

1. Explain the difference between a 5 A cable and a 13 A cable.
2. Calculate the current in a 3 kW kettle when it is connected to the 230 V mains supply.
3. **H** A mobile phone charger transfers 3600 J of energy to a phone in 20 minutes at a potential difference of 6 V. How much charge passes through the phone in that time?

Power

The power of a device tells you the rate at which the supply transfers energy to it. You calculate it with this equation:

$$P = I \times V$$

P is the power of the device in watts (W)
I is the current in the device in amperes (A)
V is the **potential difference** across the device in volts (V)

So what is the maximum power of a device that is connected to the 230 V mains supply through a 5 A fuse?

$$P = I \times V = 5\,\text{A} \times 230\,\text{V} = 1150\,\text{W or } 1.15\,\text{kW}$$

Hot wires

The wires in the cable that carry the current between the supply and a device have a small resistance. There is therefore a small potential difference across each wire, and consequently the current transfers energy to heat in the cable. The resistance of the cable has to be kept low enough to stop it getting too hot – thick cables have a lower resistance than thin ones (but cost more!).

Power can also be calculated using:

$$\underset{\text{(power in watts, W)}}{P} = \frac{E\ (\textbf{energy}\text{ in joules, J})}{t\ (\text{time in seconds, s})}$$

H Energy transfer

The energy transferred to a device can be calculated from the potential difference across it and the charge that passes through it.

$$E = V \times Q$$

E is the energy transferred to the device in joules (J)
V is the potential difference across the supply in volts (V)
Q is the charge through the device in coulombs (C)

So how much energy transfers to heat in a cable that has a potential difference of 2.3 V across it when it carries a current of 13 A for an hour?

$$I = \frac{Q}{t} = \text{so } Q = It = 13\,\text{A} \times 3600\,\text{s} = 4.68 \times 10^4\,\text{C}$$

$$E = Q \times V = 4.68 \times 10^4\,\text{C} \times 2.3\,\text{V} = 1.08 \times 10^5\,\text{J}$$

Key words

energy, potential difference, power

Working to Grade E

1 A balloon is rubbed against hair. The balloon becomes positively charged. What happens to the hair?

2 What is an electric current?

3 Draw the circuit diagram for a resistor, lamp, and switch in series with a battery.

4 Describe what this circuit diagram shows.

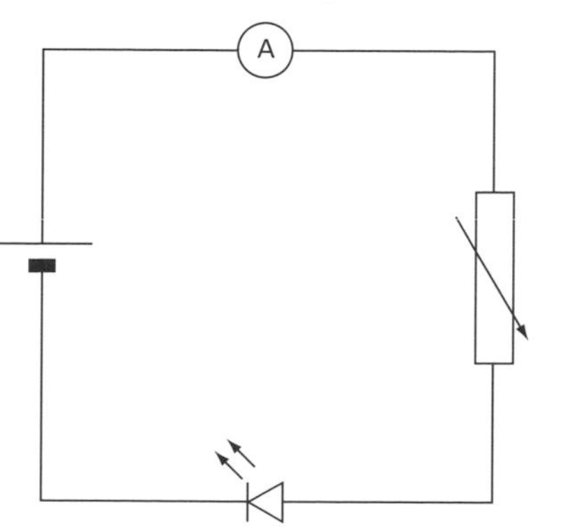

5 Calculate the potential difference across a 30 Ω resistor when it carries a current of 6 A.

6 Draw the circuit diagram for a voltmeter and resistor in parallel with a battery.

7 Name the component that can be used to detect light.

8 Complete the table for a cable connected to a UK three-pin plug.

Colour of insulation	Name of wire
blue	
	live

9 Calculate the power of a device that draws a current of 3 A from the mains supply.

Working to Grade C

10 Electrons transfer from a balloon to hair when they are rubbed together. What charge do the balloon and hair have after this transfer?

11 A charge of 1200 C flows through a TV in 10 minutes. What is the current in the TV?

12 A resistor is connected to a battery. Draw a circuit diagram to show how meters should be connected to allow the calculation of the resistance of the resistor.

13 Identify the components A, B, and C from their current–potential-difference graphs.

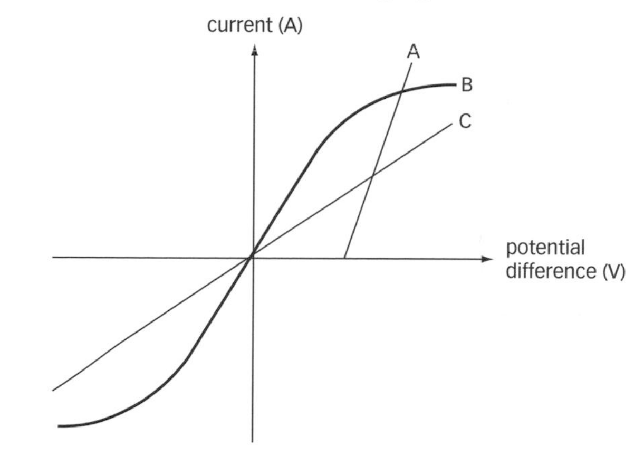

14 The current in the 12 Ω resistor is 4 A. What are the currents in the 6 Ω resistor and the battery in the diagram below?

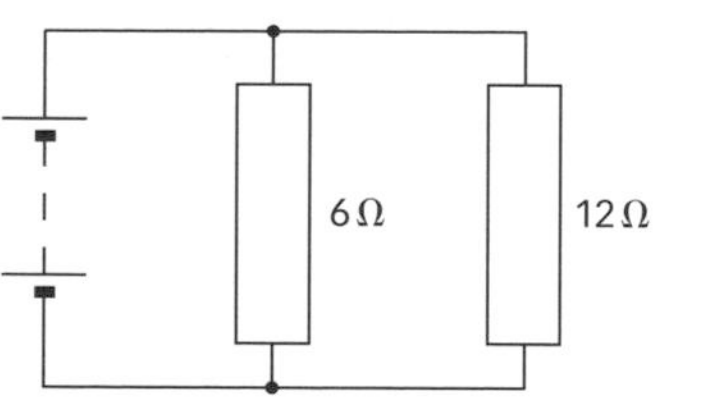

15 Explain why three-pin plugs are equipped with fuses.

Working to Grade A*

16 Calculate the maximum voltage and frequency of the a.c. power supply used for this oscilloscope trace. The timebase is set to 0.02 s/div and the vertical scale to 50 V/div.

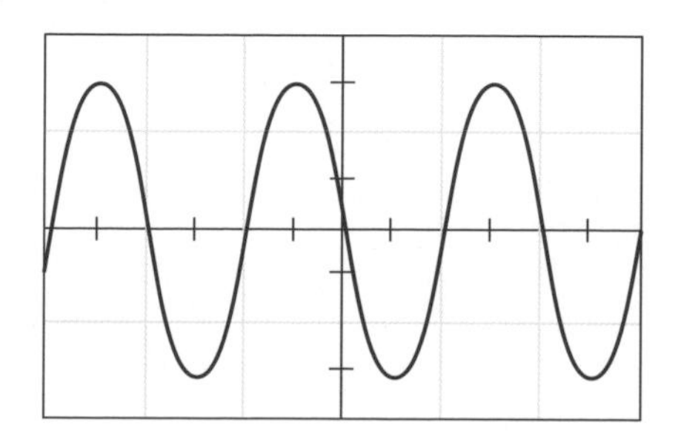

1 a Use words from the box to label the components A, B, and C in the diagram.

ammeter	**resistor**	**voltmeter**	**battery**	**switch**

A ....

C ....

B ....

(3 marks)

b The current in the 5 Ω resistor is 0.4 A.

Calculate the potential difference across the resistor.

Write down the equation you use, and then show clearly how you work out your answer.

....

....

potential difference = ... V
(2 marks)

c Plot a current–potential-difference graph for the resistor in part b on the axes below.

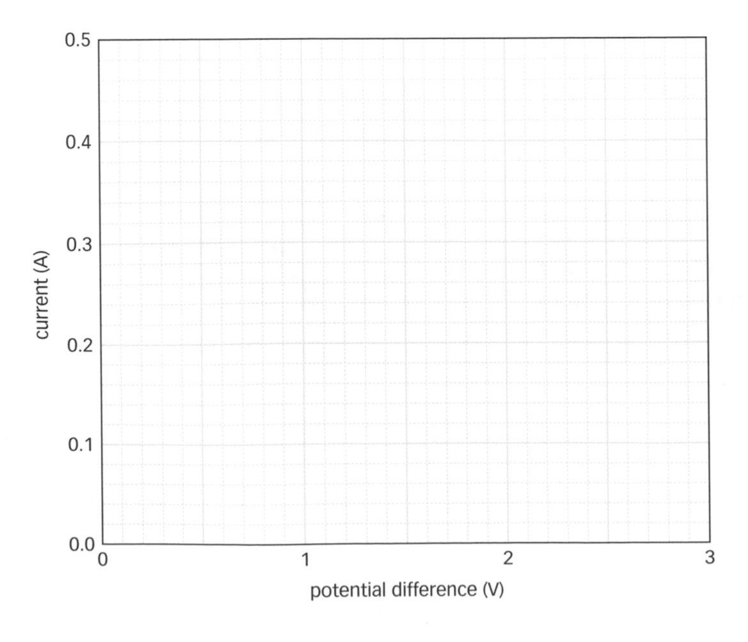

(2 marks)
(Total marks: 7)

2 A kettle is connected to the mains supply by a cable and three-pin plug.

a The statements in List A describe the function of the wires in the cable.

The colours in List B give the colours of the insulation of the wires.

Draw **one** line from each statement in List A to the correct colour in List B.

List A **Function of the wires**	**List B** **Insulation colour**
Makes the kettle safe	Red
	Blue
Carries current at 230 V	Brown
Carries low-voltage current	Green and yellow

(3 marks)

b The current in the kettle is 7 A when it is switched on.

Calculate the power of the kettle.

Write down the equation you use, and then show clearly how you work out your answer.

....

....

power = ... W

(3 marks)

c What is the best size of fuse to have in the kettle plug from part b?

Circle **one** value from the box.

1 A	3 A	5 A	13 A

(1 mark)

d Explain why there is a fuse in the plug.

....

....

....

....

(3 marks)

e When the fuse is inserted into the plug it is a conductor.

The fuse must be replaced when it becomes an insulator.

i Describe the difference between a conductor and an insulator.

....

....

....

(2 marks)

ii A student uses a cell and a lamp to test a fuse.

Draw a suitable circuit in the space below.

Use the correct symbols for each component.

(4 marks)

(Total marks: 16)

Atoms

An **atom** is the smallest part of a substance. We now believe that every atom is made from a small, heavy **nucleus** surrounded by **electrons**. The nucleus contains **protons** and **neutrons**. Most of the space in the atom is taken up by the electrons.

Each atom is specified by two quantities:

- the **atomic number** is the number of protons in the nucleus
- the **mass number** is the number of particles in the nucleus.

Isotopes

The chemical properties of an atom are determined by its electrons. Every atom has to have the same number of electrons and protons to give it an overall charge of zero. However, the number of neutrons in the nucleus is not fixed. So each element can have a number of **isotopes**, each with the same atomic number but a different mass number.

So what are the particles in an atom of the isotope $^{60}_{27}Co$?

The atomic number tells you that each atom must have 27 protons and 27 electrons. The mass number allows you to work out the number of neutrons: 60 – 27 = 33.

Atoms can gain or lose electrons to become **ions**. The charge of the ion depends on the number of electrons gained or lost.

Changing ideas

At one time scientists thought that the protons in an atom were spread evenly amongst the electrons. This was called the plum-pudding model. In 1911, Ernest Rutherford invented the modern model to explain the results of an experiment. One of his students fired alpha particles (helium nuclei) at a thin leaf of gold. Most particles passed straight through the foil, as expected, but a few bounced off it. This could not be explained by the plum-pudding model. It could only happen if each atom contained a small, heavy, charged lump – the nucleus.

Questions

1 Name the **three** different particles in each atom.

2 Calculate the numbers of each particle in an atom of $^{90}_{38}Sr$.

3 **H** How many electrons are there in a positive ion of the isotope $^{12}_{6}C$?

Revision objectives

- describe the structure of an atom in terms of neutrons, protons, and electrons
- explain the changes to an atom when it gains or loses electrons
- understand the meaning of the term isotope
- describe how new evidence caused scientists to change their model of an atom

	Relative charge	Relative mass
Proton	+1	1
Neutron	0	1
Electron	–1	0.0005

Student book references

2.27 The atom and the nucleus

Specification key

✔ P2.5.1

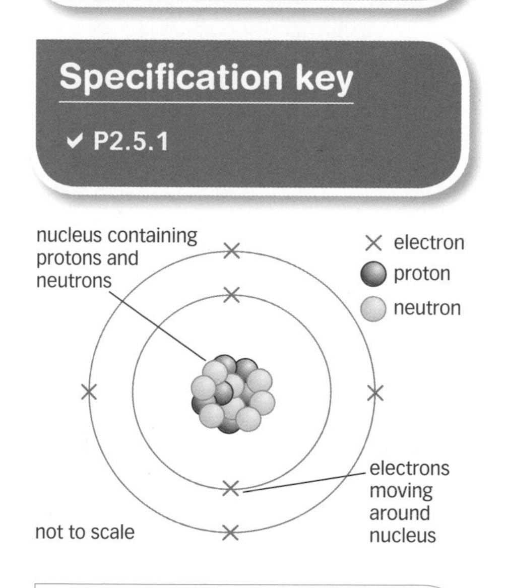

Key words

atom, nucleus, electron, proton, neutron, atomic number, mass number, isotope, ion

16: Radioactive decay

Revision objectives

- ✓ describe radioactive decay
- ✓ state sources of background radiation
- ✓ use an activity–time graph to measure half-life
- ✓ understand the effect of electric and magnetic fields on radiation
- ✓ understand the changes in the nucleus that give rise to radiation
- ✓ compare the penetrating properties of alpha, beta, and gamma radiation
- ✓ describe how to handle radioactive sources safely

Student book references

2.28 Radioactive decay and half-life

2.29 Alpha, beta, and gamma radiation

Specification key

✓ **P2.5.2 a – f, h**

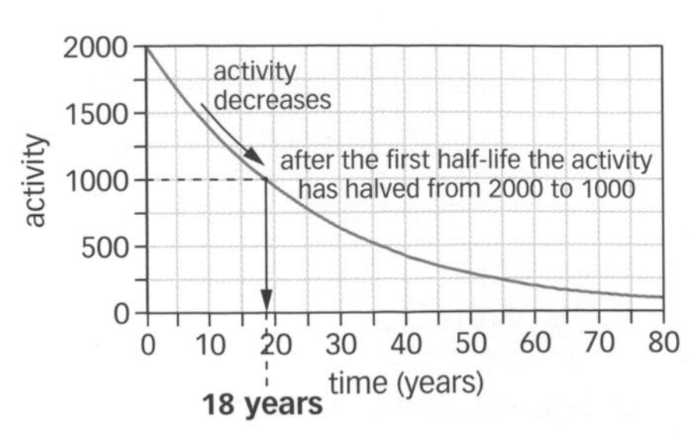

▲ The activity–time graph for a source with a half-life of 18 years.

Unstable atoms

Many isotopes are unstable. Their atoms **decay** at random, emitting three types of **ionising radiation**:

- **alpha** particles (α), each one made up of two neutrons and two protons, like a helium nucleus
- **beta** particles (β), each one made of a high-speed electron from the nucleus
- **gamma** rays (γ), each one a high-frequency electromagnetic wave.

We are exposed to this radiation all of the time. This **background radiation** comes from the **radioactive** materials in our surroundings, both natural and artificial.

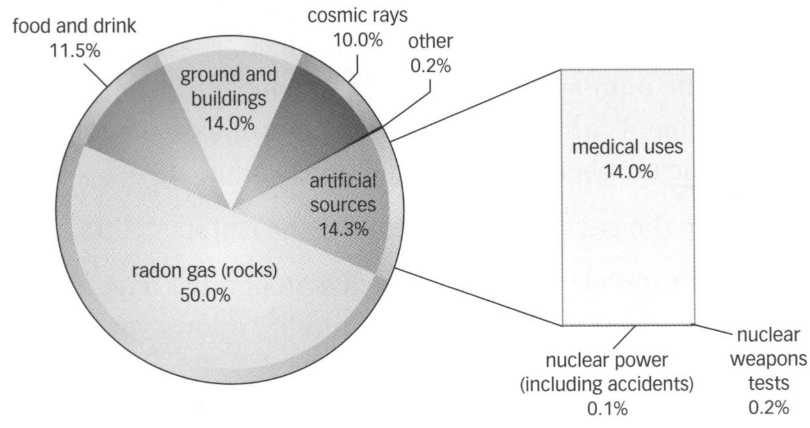

▲ Sources of background radiation in the UK.

Half-life

The radiation from radioactive materials is easy to detect because it knocks electrons out of atoms. When this **ionisation** takes place inside an insulator, the small electric current created can be detected. The output of a detector tells you about the **activity** of the source, and the rate at which its atoms decay. Since each atom can only decay once, the activity of a source decreases with time. The time taken for the activity to halve is called the **half-life**. Each radioactive isotope has a different half-life.

The half-life of cobalt-60 is 5 years. If a source has an activity of 800 today, what will the activity be in 20 years' time?

Time in years	0	5	10	15	20
Activity	800	½×800 = 400	½×400 = 200	½×200 = 100	½×100 = 50

Penetrating power

The distance that an ionising radiation can pass through matter is determined by its mass and charge. The more ions that a particle creates, the less distance it can pass through matter. Alpha particles are the most ionising because they are charged and heavy. Beta particles are less ionising because they are light. Gamma rays are the least ionising of the three because they have no charge.

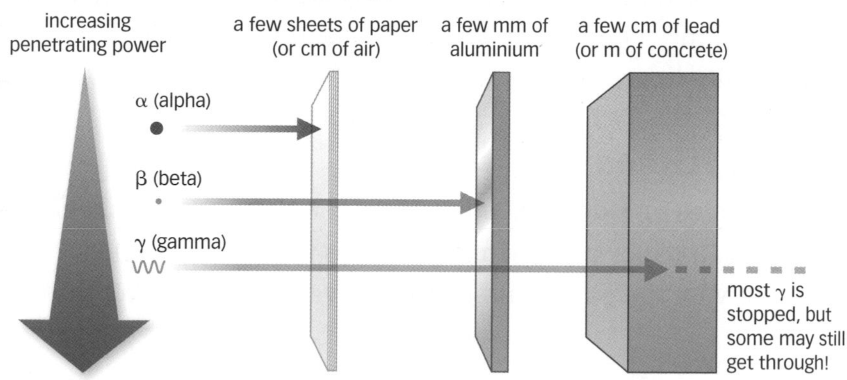

▲ Comparing the penetration of different types of radiation.

Safe handling

Ionising radiation is harmful because it ionises the cells in living tissue. This can kill cells or damage their DNA and make them cancerous. So the following precautions should be taken when handling radioactive materials to reduce your exposure:

- keep as far away as you can
- use them for the shortest time possible
- use protective clothing to stop you ingesting them
- use shielding to absorb the radiation before it gets to you.

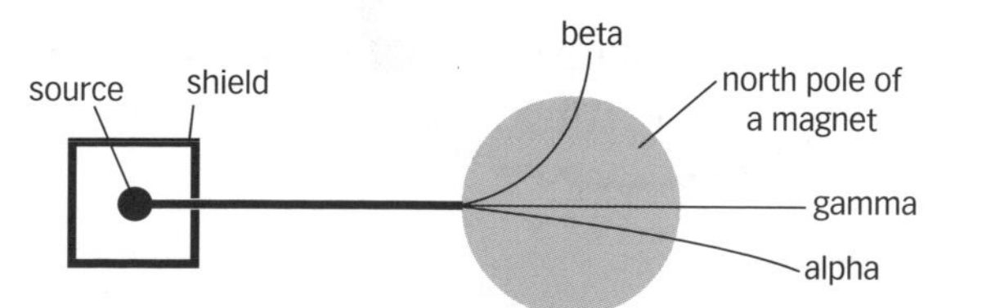

▲ The effect of a magnetic field on radiation.

Questions

1. Describe **three** safety precautions when handling radioactive materials.
2. Radiation from a source passes through a sheet of paper but is blocked by a few millimetres of aluminium. Explain the type of radiation.
3. **H** A radioactive source has an activity of 200 at 1 o'clock. By 4 o'clock the activity drops to just 25. Calculate the half-life of the source.

Key words

decay, ionising radiation, alpha, beta, gamma, background radiation, radioactive, ionisation, activity, half-life

Exam tip AQA

Write down all of the steps you go through when you do calculations with half-life.

H Nuclear equations

Equations are used to show the effect of radioactive decay on a nucleus. The equations are balanced; the mass and atomic numbers always add up to the same value on both sides of the arrow.

The emission of an alpha particle removes two neutrons and two protons from a nucleus.

$$^{222}_{86}\text{Rn} \rightarrow {}^{218}_{84}\text{Po} + {}^{4}_{2}\alpha$$

A beta particle is emitted when a neutron in the nucleus becomes a proton.

$$^{90}_{38}\text{Sr} \rightarrow {}^{90}_{39}\text{Po} + {}^{0}_{-1}\beta$$

There are no changes to the composition of the nucleus when a gamma ray is emitted.

$$^{60}_{27}\text{Co} \rightarrow {}^{60}_{27}\text{Co} + {}^{0}_{0}\gamma$$

17: Using radioactivity

Revision objectives

- describe the use of radioactive sources for detecting smoke, measuring thickness, and medical tracing
- describe the processes of nuclear fusion and fission
- compare the advantages and disadvantages of using fission and fusion to produce energy

Student book references

2.30 Uses of radiation

2.31 Fission and fusion

Specification key

✔ P2.5.2 g ✔ P2.6.1

✔ P2.6.2 a – b

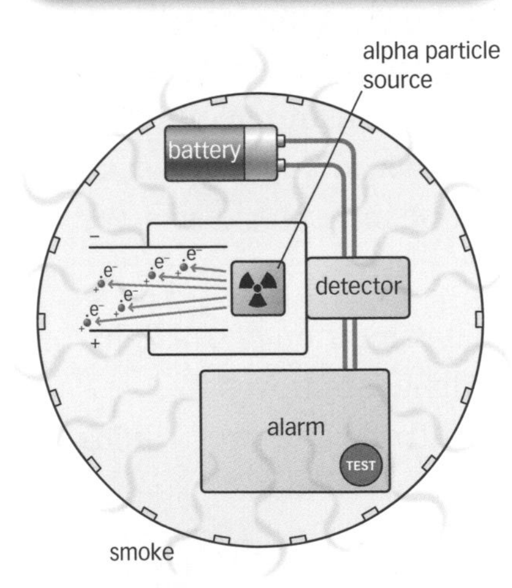

▲ A smoke detector.

Alpha particles

The smoke detector below contains an alpha particle source. Alpha particles ionise the air in the chamber next to the source, turning the air from an insulator into a conductor. So a potential difference across the chamber results in a current that can be detected. Any smoke that enters the chamber reduces the ionisation of the air. The resulting drop in current is detected and triggers the alarm. The source needs to have a long half-life to avoid false alarms!

Beta particles

Beta particles are ideal for measuring the thickness of sheets of paper in a paper mill. Alpha particles would be stopped completely by the paper and gamma rays would go straight through it.

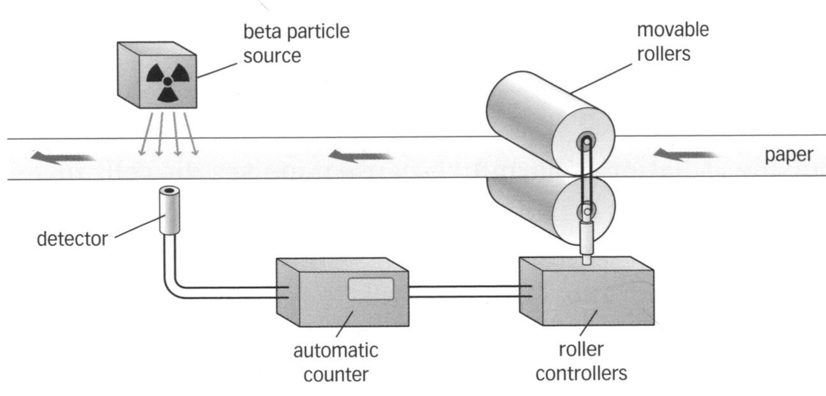

▲ Automatic thickness control of paper.

Any increase in thickness of the paper reduces the rate at which beta particles reach the detector. This can be fed back to the rollers to reduce the thickness back to the target value.

Gamma rays

Medical **tracers** containing gamma ray sources are used to create images of organs inside the body. This is because only gamma rays are easily able to pass through the body. The tracer is injected into the body and gamma ray detectors around the body record where the tracer goes. The tracer needs to have a short half-life so that the patient is only radiaoctive for a few days!

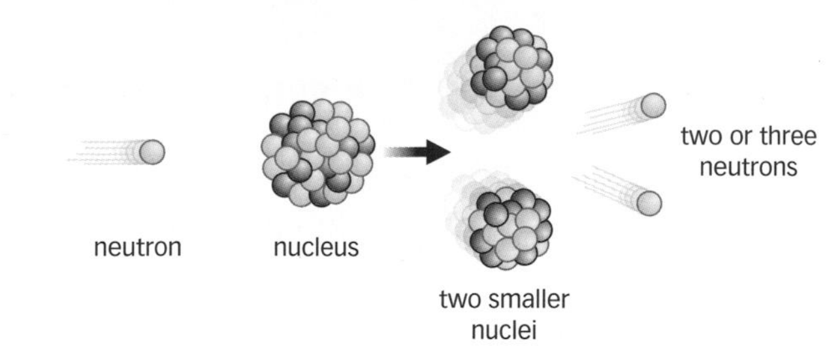

▲ When the nucleus absorbs a neutron it splits into two.

Fission

Two radioactive materials (uranium-235 or plutonium-239) can be used to produce energy in a nuclear reactor. They split (**fission**) into two smaller nuclei when they absorb a neutron. This releases a lot of energy as heat, which can be used to boil water to steam and spin a turbine. Each fission produces two or more neutrons, each of which can go on to produce the fission of another nucleus. If this **chain reaction** is carefully controlled it can result in a steady release of energy.

Fusion

Nuclear reactors use the fission of a large nucleus into two smaller ones to release energy. Stars release energy through the **fusion** of two small nuclei into larger nuclei. The Sun fuses four hydrogen nuclei to make a single nucleus of helium. It has done this for billions of years, transferring most of the energy released to the rest of the Universe as light. The nuclei that fuse have a positive charge, so they repel each other. A very high temperature is required to give the particles enough energy to get close enough for fusion to take place. This is proving difficult to achieve on Earth.

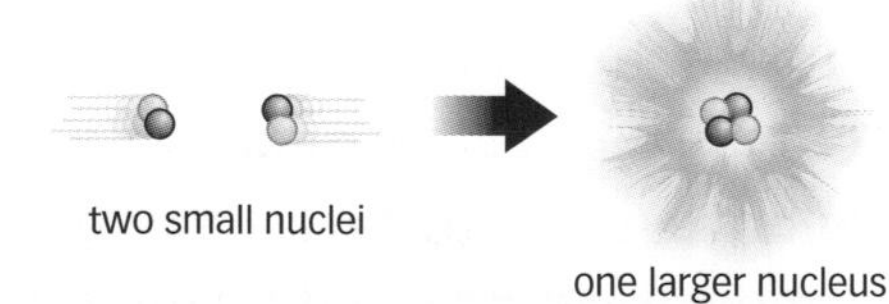

▲ The fusion of two small nuclei can release energy.

Questions

1 Name **two** substances that are used as fuel in a fission reactor.

2 Explain why a medical tracer should be a gamma ray source with a short half-life.

3 **H** Explain why a long-lived beta source is used to control the thickness of paper in a paper mill.

Key words

tracers, fission, chain reaction, fusion

Exam tip

AQA

Use the properties of a type of radiation to justify its use in an application.

18: Star lives

Revision objectives

- explain the formation of stars
- describe what happens when a star approaches the end of its life
- describe the complete life cycle of a star including the production of new elements

Student book references

2.32 Star life cycles

Specification key

- **P2.6.2 c – f**

Key words

nebula, protostar, star, fusion, main sequence, red giant, white dwarf, super red giant, supernova, neutron star, black hole

Questions

1. Here are some stages in a star's life. Put them in the correct order, starting with the earliest: protostar, nebula, white dwarf, red giant star.
2. Explain how stars form in space.
3. **H** Explain why a star is only stable during its main sequence.

Birth

All new stars are created the same way.

- A large cloud of gas called a **nebula** starts to collapse under its own weight.
- Gravity pulls the particles (mostly hydrogen) closer to each other, heating the cloud.
- A glowing hot ball of gas called a **protostar** forms at the centre.
- The protostar gets hotter as more material falls into it.
- A **star** forms when the temperature at the centre is high enough for **fusion** reactions.

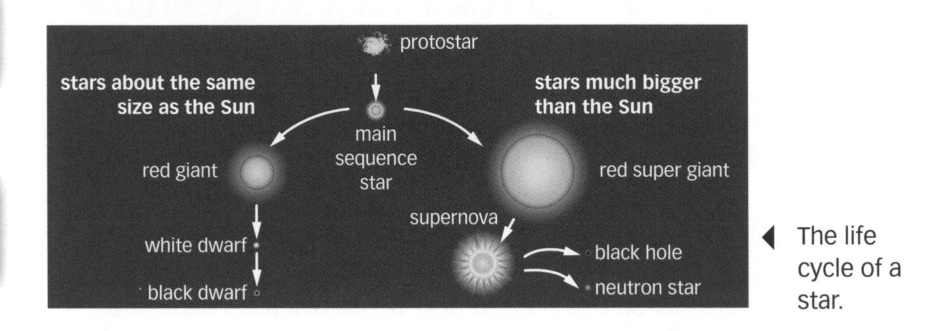

The life cycle of a star.

Death

The energy released by the fusion of hydrogen nuclei into helium stops a star from collapsing. The outward push of the energy flowing to the surface balances the inwards push of gravity. This balance keeps the star stable. This **main sequence** part of the star's life ends when it runs out of hydrogen fuel at its centre. Small stars, like our Sun, last several billion years before swelling up into a **red giant** and then cooling to a **white dwarf** as fusion reactions stop. Large stars have shorter main sequences and turn into **red super giants**. These can fuse helium nuclei into heavier elements. A heavy enough star can convert its nuclei into iron before it explodes as a **supernova**, creating new elements (and elements heavier than iron) and spreading them through space. All of the elements on Earth, apart from hydrogen and helium, were made this way. The material left behind after a supernova collapses under its own weight until it forms a dense **neutron star**. Sometimes a neutron star is heavy enough to collapse forever, becoming a **black hole**.

Exam tip

AQA

Remember that what happens at the end of its main sequence depends upon the mass of a star.

Working to Grade E

1 Complete this table for the particles in an atom.

Name	Mass	Charge
	0	–1
	1	1
		0

2 Name the **two** particles in a nucleus.

3 State **three** sources of background radiation.

4 Name the radiation that is most easily stopped.

5 State **three** precautions to take when handling radioactive materials.

6 Name the best type of radioactive source for a medical tracer.

7 Complete these sentences for a nuclear reactor.

The nuclei of the ________ fuel split into two when they absorb a ________. This is called nuclear ________ .

8 Put these stages of star formation in the correct order, ending with the last: black hole, main sequence, protostar, supernova, super red giant.

Working to Grade C

9 Calculate the number of electrons, protons, and neutrons in one atom of $^{22}_{10}$Ne.

10 State the similarities and differences between the isotopes $^{22}_{10}$Ne and $^{20}_{10}$Ne.

11 A source has a half-life of 5 hours. If its activity is 1000 now, what will it be in 10 hours' time?

12 Explain why radioactive sources can be harmful.

13 Explain **three** different precautions to take when handling radioactive materials.

14 State which types of radiation can pass through paper.

15 Explain why a medical tracer should be a gamma ray emitter with a half-life of a few hours.

16 Explain the stages in the life of a star like the Sun.

Working to Grade A*

17 The activity of a source drops from 512 to 32 in just 8 hours. What is its half-life?

18 Complete the symbol $^{_}_{_}$Np for the product of the alpha decay of $^{241}_{95}$Am.

19 Explain the changes in a nucleus when it emits a beta particle.

Examination questions
Radioactivity

1 The table shows the average radiation dose that a person in various places in the UK receives from background radiation in a year.

The doses are measured in millisieverts (mSv).

UK location	Annual dose in mSv
Cornwall	9
East Anglia	1
London	2

a Give **two** sources of background radiation.

1 ....

2 ....

(2 marks)

b Which of these is the correct reason why background radiation is harmful?

Tick (✓) **one** box.

It is in all of the food that you eat. ☐

It damages cells by ionising them. ☐

It can't be detected without instruments. ☐

It contains three different sorts of radiation. ☐

(1 mark)

c A radiographer works with radiation in a London hospital.

She is only allowed to receive a total dose of 20 mSv of radiation in a year.

Her average dose from the radiation at work is 0.02 mSv per hour.

Calculate the number of hours she is allowed to work each year.

....

....

(3 marks)

(Total marks: 6)

2 This diagram represents an atom of the element lithium.

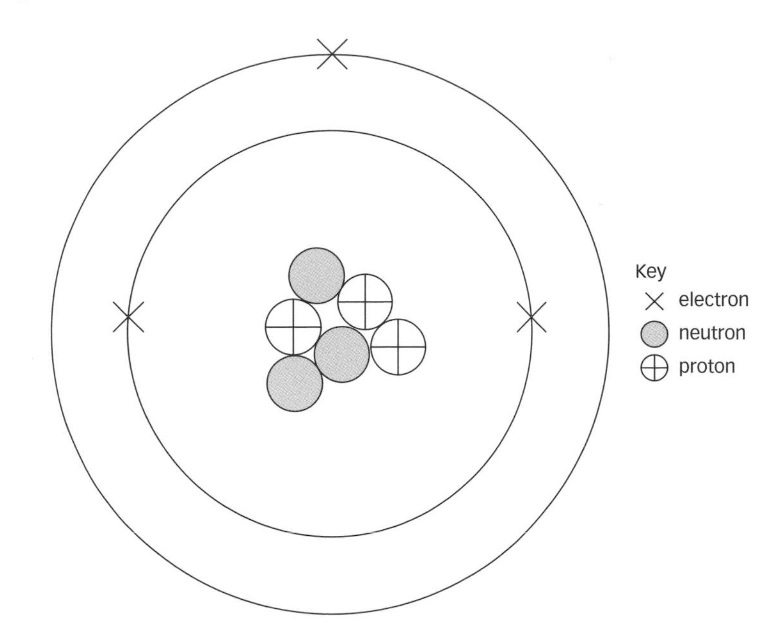

a Complete the symbol for this isotope of lithium.

$^{\cdots\cdots\cdots}_{\cdots\cdots\cdots}\text{Li}$

(2 marks)

b Another isotope of lithium has a mass number of 7. It is called lithium-7.

Complete the diagram below to represent an atom of this isotope.

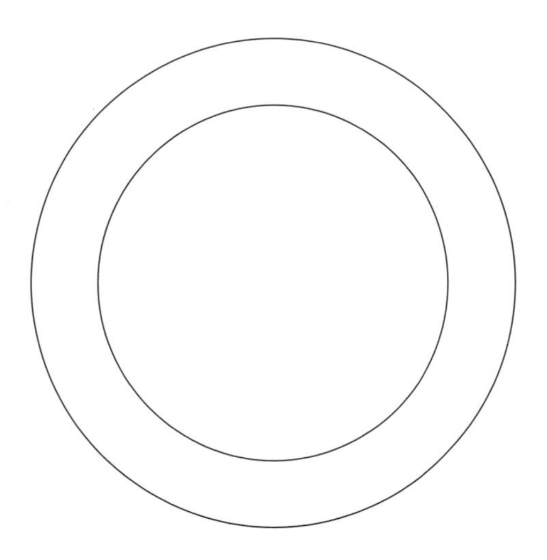

(3 marks)

c When lithium-7 absorbs a neutron it breaks into a pair of alpha particles.

Use a word from the box to complete the sentence.

fission	**fusion**	**radioactive**	**star**

This is an example of a .... reaction.

(1 mark)

d A smoke detector contains a radioactive source.

Radiation from the source ionises air in the detector.

Any smoke in the detector reduces this ionisation.

Explain why the source needs to:
- emit alpha particles
- have a long half-life.

....

....

....

....

(3 marks)

e The people who assemble smoke detectors need to take safety precautions.

Describe and explain **one** precaution that they should take.

....

....

....

(2 marks)

f List A contains the names of three different types of radiation.

List B describes these types of radiation.

Draw **one** line from each name in list A to its description in list B.

List A **Types of radiation**	**List B** **Description**
	Helium nucleus
Beta	
	Hydrogen nucleus
Alpha	
	High-speed electron
Gamma	
	Electromagnetic wave

(3 marks)
(Total marks: 14)

3 a This diagram shows stages in the life cycle of a star.

Use words from the box below to complete the diagram.

main sequence	**neutron star** **supernova**	**protostar** **white dwarf**	**super red giant**

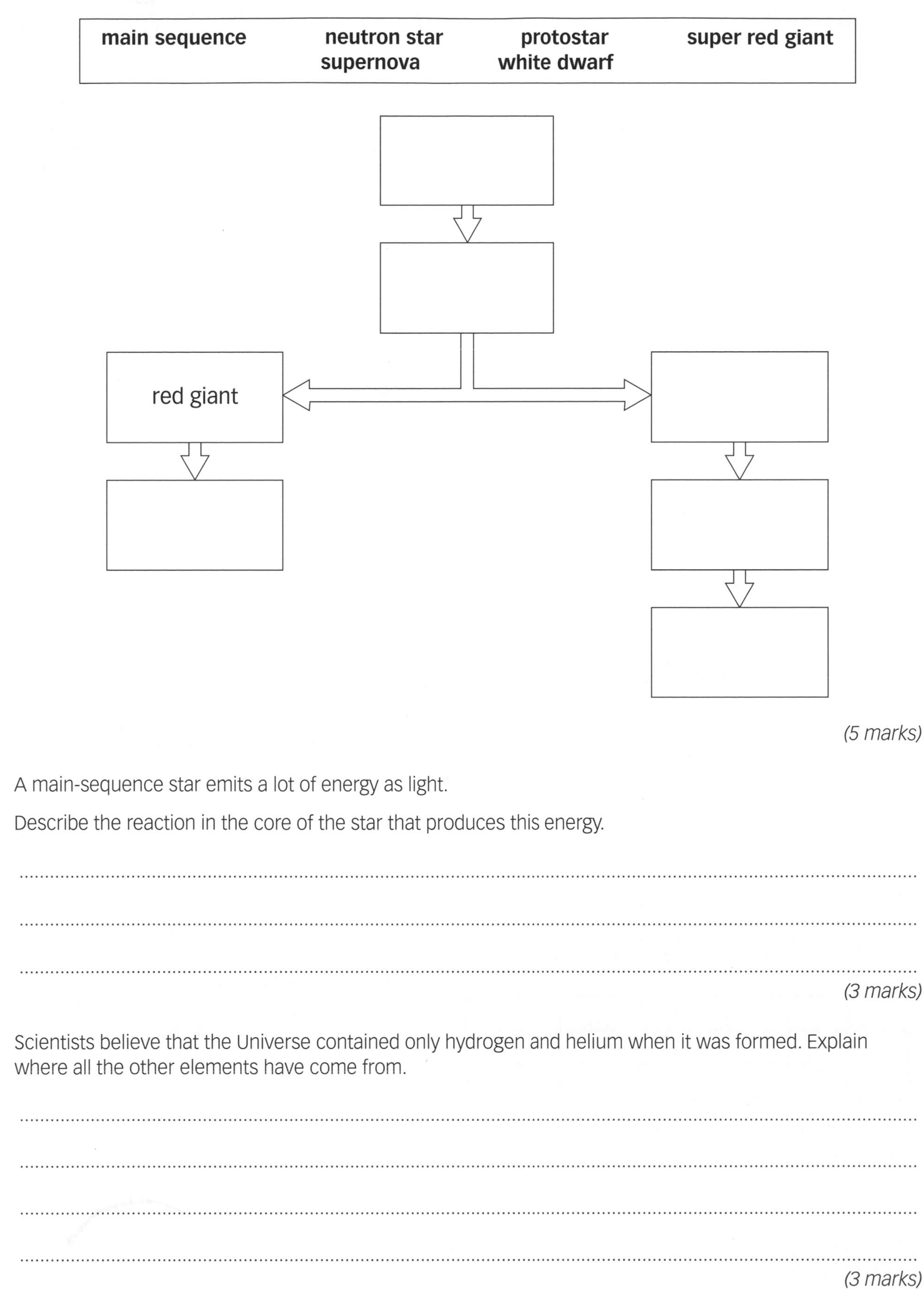

(5 marks)

b A main-sequence star emits a lot of energy as light.

Describe the reaction in the core of the star that produces this energy.

....

....

....

(3 marks)

c Scientists believe that the Universe contained only hydrogen and helium when it was formed. Explain where all the other elements have come from.

....

....

....

....

(3 marks)

(Total marks: 11)

Designing an investigation and making measurements

In this module there are some opportunities to design investigations and make measurements. These include measurements of half-life and the penetrating power of radiation.

Although you are unlikely to demonstrate your investigative skills practically, you are likely to be asked to comment on investigations done by others. The example below offers guidance in this skill area. It also gives you the chance to practise using your skills to answer the sorts of questions that may well come up in exams.

Comparing the usefulness of different thicknesses of shielding

Skill – Understanding the experiment

Sue tested the hypothesis that doubling the thickness of lead shielding halves the rate at which gamma rays get through it.

She placed the radioactive source near a detector, with the lead sheets in between. For each lead sheet in place, she recorded how many gamma rays were detected in each minute. Three measurements were made for each thickness of lead.

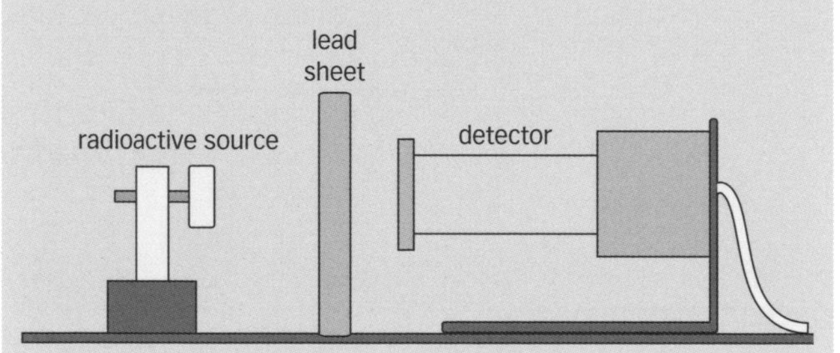

Sue's results are in the table below.

Lead sheet thickness (mm)	Gamma rays (per minute)		
	First trial	Second trial	Third trial
1.0	100	111	107
2.0	79	84	90
4.0	47	60	52
8.0	20	17	23

1 Identify the independent and dependent variables.

In an investigation:

- the independent variables are the ones that are changed by the scientist
- the dependent variable is the one that is measured for each change of the independent variables.

2 Suggest an important control variable for the investigation.

A control variable is one that the investigator thinks might affect the outcome, so they try to keep this variable the same all the way through.

Skill – Using data to draw conclusions

3 How should Sue combine the data from the three trials?

The best way of combining the data from different measurements of the dependent variable is to calculate the average.

4 How can Sue use the data to test her hypothesis?

Each sheet of lead has double the thickness of the previous one. So Sue should divide the average detector reading by the previous one. If the answer is always 0.5, then her hypothesis is correct. A different answer that is still the same for all thicknesses would suggest that her hypothesis could be correct for a different thickness. She could plot a scatter graph of the detector reading against the shielding thickness to find out if there is still a halving law, but not for the thicknesses that she has chosen.

Skill – Evaluating the experiment

5 What should Sue do about the large variation in results for each setting of the independent variable?

The range of data is often reduced by making more measurements. She could do this by either counting gamma rays for more than a minute, or making more measurements for a minute each time (more trials).

AQA Upgrade

Answering an extended question

QUESTION

In this question you will be assessed on using good English, organising information clearly, and using specialist terms where appropriate.

1 All stars go through a life cycle. Describe and explain the life cycle of a star that is much heavier than the Sun. *(6 marks)*

G–E

Well, the star is exploding, then vanish into a black dwarf, because it is too hevy to get away from itself. It starts off a gas which gets hotter and hotter then turns into a star when the gas starts to burn. It explodes when it runs out of fule.

Examiner: This answer only earns 1 mark, so is typical of an F-grade candidate. The spelling and grammar are poor, information is presented in a confusing order, and only some of the science is correct. There is incorrect use of a technical term: 'fusion' should be used instead of 'burn'.

D–C

The star begins as a protostar, this is a gas which gets smaller and smaller and hotter and hotter until the hydrogen starts to turn into a helium. This is main sequence and lasts for a long time until all of the hydrogen is gone. The star grows into a red giant and then explodes when it has a core of iron. Lots of different elements are made in the explosion, these get spread around and eventually get pulled together to make another star, only this time with planets around it.

Examiner: This answer earns 3 marks, typical of a C-grade candidate. It contains a lot of correct information, presented in a logical order with correct spelling. Some sentences are too long. There is little information about fusion and the fate of the star after it has become a supernova, and is mostly a succession of facts with very little attempt at explanation.

B–A*

A cloud of hydrogen gas shrinks as gravity pulls all of its particles towards each other. This transfers GPE to heat, so as the cloud shrinks its temperature increases until the centre is hot enough for hydrogen nuclei to fuse into helium nuclei. This process releases a lot of energy as gamma rays. The outwards flow of this energy stops the star from shrinking any more. When all of the hydrogen in the core has been converted to helium, it contracts until the helium can fuse into heavier elements. The outer layers of the star swell up to form a super red giant. When the core is solid iron, fusion stops and so it collapses under its own weight to make a black hole. The outer layers explode as a supernova, fusing nuclei together to make elements heavier than iron.

Examiner: This answer earns 6 marks, typical of an A*-grade candidate. Each step in the evolution of a star has been correctly named, described, and explained. Information is presented in a logical order (start to finish), with no errors of grammar or spelling.

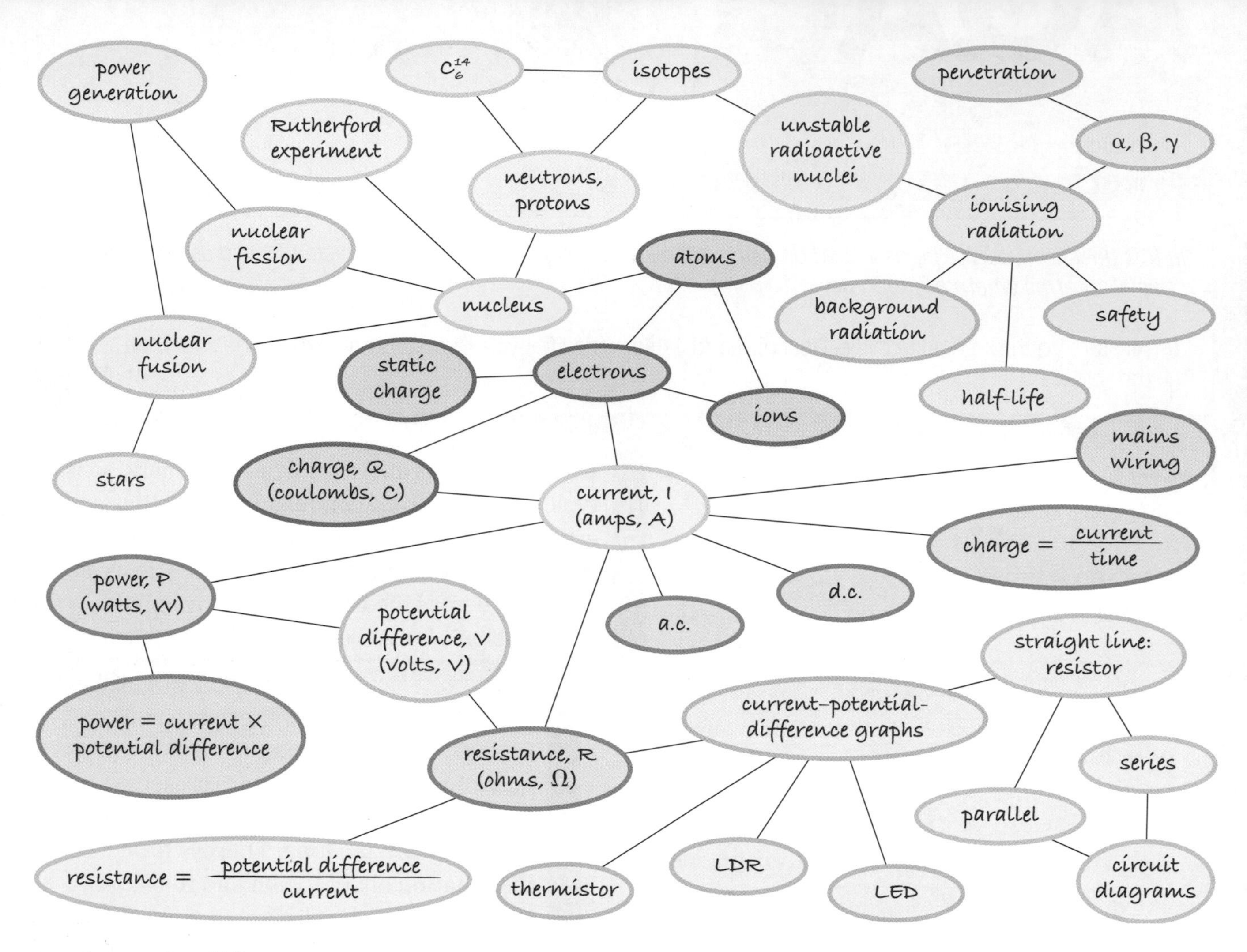

Revision checklist

- Some materials gain an electrostatic charge when rubbed together. Opposite charges attract, like repel.
- Electric current is the flow of charge around a circuit.
- The potential difference between two points in an electronic circuit is the work done (energy transferred) per coulomb of charge that passes through the points.
- Circuit symbols represent electrical components.
- Resistance is a measure of how difficult it is for electrons to pass through a component.
- Current must pass through all components in a series circuit one after the other. Current in a parallel circuit can take more than one route.
- The resistance of a filament lamp increases as its temperature increases.
- A light-emitting diode (LED) produces light when a current flows through it, and is efficient and long-lasting.
- The resistance of a light-dependent resistor (LDR) varies with intensity of light. LDRs are used in switches.
- The resistance of a thermistor varies as its temperature changes. Thermistors are used in temperature sensors.
- Electric current has two forms: direct current (d.c.), usually produced by batteries or cells; and alternating current (a.c.), such as UK mains electricity (230 V a.c. with frequency 50 Hz).
- UK three-pin plugs contain a live wire (brown), a neutral wire (blue), an earth wire (green and yellow), and a fuse. Fuses, earth wires, and circuit breakers are safety devices.
- When an electrical charge flows through a resistor, energy is transferred. Larger devices need larger power cables. The power of a device is its rate of energy transfer.
- Atoms have a small central nucleus of protons (positively charged) and neutrons (no charge) surrounded by negatively charged electrons in orbit. Positively or negatively charged ions are created when atoms gain or lose electrons.
- Isotopes are atoms with the same number of protons (same atomic number) but a different number of neutrons.
- Radioactive decay is the breakdown of the nucleus of an atom to form ionising radiation. Half-life is the average time taken for activity (decays per second) to halve.
- Nuclear fission is used in nuclear reactors to produce electricity.
- Nuclear fusion produces energy in stars.

Answers

B2 1: Cells and cell structure

1 An organelle/structure in the cell of plants and animals that contains the chromosomes. It controls the function of the cell.
2 Plant cells have: cell wall, cell membrane, cytoplasm, chloroplasts, vacuole, mitochondria, ribosomes, and nucleus. An animal cell does not have a cell wall, a vacuole, or chloroplasts.
3 Differentiation is the process whereby cells become specialised to do a particular job.

B2 2: Diffusion

1 Diffusion is the way molecules get into and out of cells, and this is necessary for cells to work.
2 Because particles don't move around in a solid.
3 The greater the surface area, the faster the rate of diffusion.

B2 1–2 Levelled questions: Cells and diffusion

Working to Grade E

1 a i A
ii B
iii C
b Chloroplasts, cell wall, and vacuole.
2 Inside the permanent vacuole.

Working to Grade C

3 a Controls the activities of the cell.
b Supports the cell.
c Controls the movement of substances into and out of the cell.
d Traps light energy for photosynthesis.
4 a Nerve cell – long extension; muscle cell – contractile proteins; palisade cell – chloroplasts.
b Nerve cell – takes impulses to the brain; muscle cell – contracts to allow the cell to shorten; palisade cell – allows photosynthesis.
5 Differentiation
6

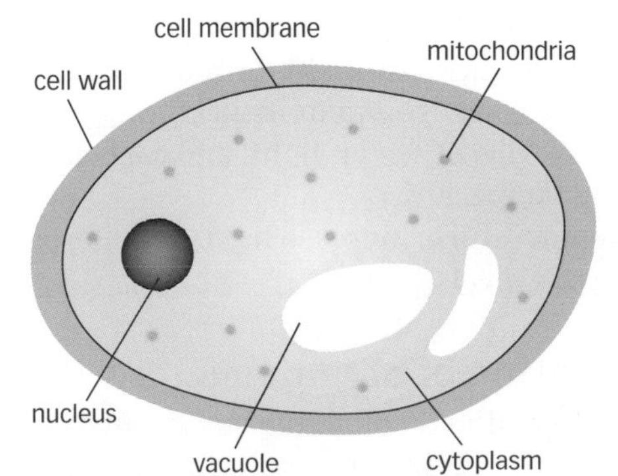

7 They don't have chloroplasts.
8 The net movement of particles from an area of high concentration to an area of low concentration, until the concentrations even out.
9 Oxygen
10 Distance, concentration gradient, and surface area.

Working to Grade A*

11 Make proteins
12 Cell wall is made of a different chemical; there is no distinct nucleus.
13 There is no diffusion occurring at A and C. At B, diffusion occurs into the cell. At D, diffusion occurs out of the cell.
14 This is the difference in concentration between the two areas.
15 a This is how fast diffusion occurs.
b Distance – the shorter the distance the particles have to move, the faster the rate.
Concentration gradient – the greater the difference in concentration, the faster the rate.
Surface area – the greater the surface area, the faster the rate.
16 The lungs have a large surface area for the movement of oxygen. Therefore there is more surface area over which the oxygen molecules can move.

B2 1–2 Examination questions: Cells and diffusion

1 a 1 mark will be awarded for each correct structure/function completed in the table, up to a total of 8.

Structure	Function
Cell membrane	**Controls the movement of substances into and out of the cell.**
Nucleus	Controls the activities of the cell, contains DNA.
Cell wall	**Strong structure that supports the cell.**
Mitochondria	Releases energy from sugar during aerobic respiration.
Chloroplasts	**(Contains chlorophyll which traps light) to carry out photosynthesis.**
Vacuole	**Stores liquid (cell sap) used for support.**
Cytoplasm	Where many chemical reactions occur.
Ribosomes	Proteins are made here.

b 1 mark will be awarded for each of the following three structures: chloroplast; cell wall; vacuole.
2 a The net movement of particles (1) from an area of high concentration to an area of low concentration (1).
b Oxygen (1)

B2 3: Animal tissues and organs

1 When an organism is built of many cells.
2 Three from: mouth, oesophagus, stomach, small intestine, large intestine, pancreas, or liver.
3 A tissue is made of more or less similar cells working together, whereas an organ is a group of different tissues working together.

B2 4: Plant tissues and organs

1 Three from: epidermal tissue, mesophyll, xylem and phloem.
2 Phloem and xylem.
3 They are made of hollow cells with strong cell walls, which are stacked to form a long tube through the plant.

B2 3–4 Levelled questions: Tissues and organs

Working to Grade E

1 A group of similar cells working together.
2 a Muscle, glandular tissue, epithelial tissue, or any other reasonable answer.
  b Heart, stomach, or any other reasonable answer.
3 a Reproductive (A), circulatory (B), skeletal (C).
  b Heart
4 A group of different tissues working together at a specific function.
5 a Mouth
  b Large intestine
  c Small intestine
6 A: mouth, B: oesophagus, C: liver, D: stomach, E: pancreas, F: small intestine, G: large intestine, H: rectum.
7 a Supports the plant and transports molecules through the plant.
  b Production of food by photosynthesis.
  c Anchors the plant, uptake of water and minerals from the soil.

Working to Grade C

8 a Reproductive: reproduction; circulatory: transport; skeletal: support and movement.
  b As they are each made of a number of different organs working together at a specific function.
9 a It contracts causing the stomach to move and churn up the food.
  b Glandular tissue
  c On the outside and inside lining of the stomach.
10 a Produce a digestive juice (containing enzymes).
  b Produces bile, which aids digestion.
  c These produce a digestive juice, which is added into the mouth.
11 It is made of several organs working together at a specific function.
12 Epidermal tissue
13 Water
14

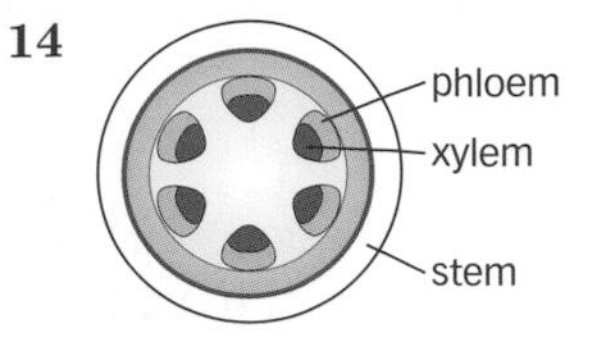

15 Stacked one above the other to form a long tube.
16 Inside the leaf.
17 Photosynthesis
18 Sugars

Working to Grade A*

19 Each tissue has a different role in the stomach. For the complete function of the stomach, all three roles are needed.
20 a The leaves
  b Other parts of the plant.
21 a Xylem
  b They have strong cell walls.

B2 3–4 Examination questions: Tissues and organs

1 a Answer should include any five of:
- epithelium on the outside of stomach
- covers the outside, protects the stomach
- muscular tissue in wall
- can contract and churn the contents up
- glandular tissue on inside
- produces acid and enzymes to help digest food
- inner epithelium
- to line the stomach.

1 mark will be awarded for each up to a total of 5.

B2 5: Photosynthesis

1 To make food for the plant.
2 Light energy into chemical energy.
3 Water is absorbed from the soil by the roots, and it moves through the xylem to the leaf. Carbon dioxide diffuses through pores in the leaf and into the mesophyll cells.

B2 6: Rates of photosynthesis

1 Because there is more sunlight for photosynthesis.
2 They control the environment inside the greenhouse (levels of water, light, and carbon dioxide, and temperature).
3 Increase temperature, increase light, and increase carbon dioxide levels.

B2 7: Distribution of organisms

1 Place a transect line through the environment and use a quadrat at regular intervals to count the number of organisms.
2 Mean, median, and mode.
3 A combination of two things: the availability/ suitability of key factors such as temperature, light, carbon dioxide, nutrients, oxygen, and water; and the organism's adaptations to cope with the conditions.

B2 5–7 Levelled questions: Photosynthesis and distribution

Working to Grade E

1 Carbon dioxide and water.
2 Glucose/carbohydrates/sugars/starch and oxygen.
3 Sunlight
4 Chlorophyll
5 Leaf
6 Two from: plants, (some bacteria), and algae.
7 a A = upper epidermis; B = palisade mesophyll layer; C = spongy mesophyll layer; D = lower epidermis.
  b Palisade mesophyll layer.
8 The number of individuals of a species in a given area.
9 a In general, as the light increases the numbers of plants increase. But very high light levels will reduce plant numbers.
  b The more light, the more photosynthesis. But very high levels tend to increase temperature.
10 To count the numbers and types of organism in an area.
11 Polar bears live in colder environments.
12 All of the plant and animal populations in an area.

Working to Grade C

13 Carbon dioxide + water → glucose + oxygen
14 Through the stoma on the lower epidermis.
15 In a process that is controlled by a number of factors, the limiting factor is the factor which is at the lowest level and limits the rate of reaction.
16 The speed at which photosynthesis occurs.
17 a Carbon dioxide
  b No, it is now likely to be temperature.
18 a i Heat and carbon dioxide.
    ii They will increase the rate of photosynthesis (as long as there is plenty of light), and the glucose can be used for growth.
  b Stops the greenhouse getting too hot and killing the plants.
  c So that water can drain out of the pots to prevent rotting of roots.
19 It is very expensive, so reduces profit.
20 The temperature is hot in the day, and it is very dry.
21 a To make the results more reliable.
  b The readings were taken at set regular intervals that they could not change.
  c The number of daisies increases as you move further from the school.
  d The areas close to the school buildings have more shade.
22 a The bison eat the grassland plants, so high numbers of bison means a lower number of grass plants. So the bison need to move to an area with a fresh supply of grass plants.
  b The bison will either die or move on.
23 The mode is the most common value in a set of data. The median is the middle value when the data is in rank order.
24 a The larger the sample size, the more valid the results. This makes the data more accurate.
  b If results cannot be repeated then the conclusions may not be valid.

Working to Grade A*

25 a Arrow should pass though the stoma, through the air spaces up and into a palisade cell.
  b Sucrose is transported around the plant in the phloem. (Oxygen is released through the stoma.)
26 a i Nitrogen
    ii Nitrate ions from the soil.
  b Cellulose.
  c To store food and for growth.
  d Sucrose
27 Glucose
28 Because it is not soluble and so will not dissolve and leave the cell in water.
29 a A
  b As the light increases the rate of photosynthesis increases.
  c Another factor becomes limiting.
30 a On a warm day with plenty of sunshine and high traffic levels.
  b Because temperature and light levels are high, and the traffic gives off carbon dioxide, increasing its levels.
31 a Increased by electric lighting; decreased by netting or whitewash.
  b Maximum light will lead to a higher rate of photosynthesis. Too much light might raise the temperature too high and kill the plants.
32 Strawberries can be grown for a longer period during the year.
33 Light readings over a 24-hour period.

B2 5–7 Examination questions: Photosynthesis and distribution

1 1 mark will be awarded for each correct label, up to a total of 6.

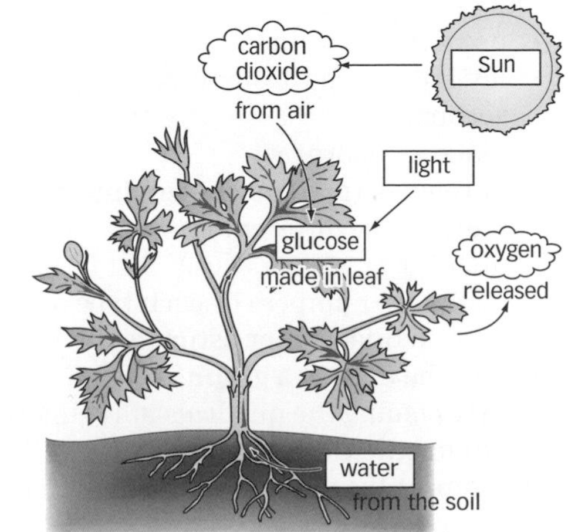

2 a You will gain 2 marks for plotting the points correctly and 1 mark for drawing the line correctly. 1 mark will be deducted for each plotting error up to a maximum of –2.

b 32.5 (+/–1) (1)

c 1 mark will be awarded for each of the following points, up to a maximum of 3.

- As the light intensity increases, the rate of photosynthesis increases.
- This is because more light energy can be trapped by the chlorophyll and used to build the sugars in photosynthesis.
- However, eventually there is no increase in the rate, despite the increase in light, probably due to other factors limiting the rate.

d 2 marks to be awarded as follows: When a process is affected by several factors, the one that is at the lowest level (1), will limit the rate of reaction (1).

e 1 mark for each of temperature and carbon dioxide.

3 a i Tube A (1)

ii 1 mark for each of the following parts of the answer, up to a total of 3:

- because the indicator has turned purple
- as the leaf has used up all of the carbon dioxide
- for photosynthesis.

b To keep the conditions the same in the two tests/to carry out a fair test. (1)

c Humans breathe out carbon dioxide (1), which will affect the indicator solution colour (1).

d i 1 mark for identifying that different amounts of carbon dioxide were absorbed (because the size of the leaf was different) and 1 mark for identifying that this was due to different amounts of photosynthesis (more carbon dioxide was absorbed by the bigger leaf in A, and more photosynthesis is able to occur in bigger leaves).

ii There would be no photosynthesis, as the leaf would be dead, so there would be no change in indicator colour. (1)

4 a Any 5 from the following list, up to a total of 5 marks:

- use a transect line
- lay it through the habitat, e.g. down a seashore
- place a quadrat
- at regular (predetermined) intervals
- identify the different seaweeds in each quadrat
- count the numbers of each type of seaweed in each quadrat (or estimate the % cover)
- plot the data as a graph.

b i By placing the quadrats at regular points along the transect. (1)

ii Repeat the experiment in the same location again. (1)

How Science Works: Cells and the growing plant

1 1 mark will be awarded for identifying that Dog's mercury will grow better at high light intensities. 1 mark will be awarded for a scientific explanation that where there is more light, there will be a higher rate of photosynthesis, making food for the plant to grow.

2 a 1 mark will be awarded for a biological variable from your hypothesis, such as the number of Dog's mercury plants or percentage cover.
1 mark will be awarded for the physical factor from your hypothesis, such as the light intensity.

b Refer to the guidance given in the Student Book.
1 mark will be awarded for outlining the correct use of apparatus; 1 mark will be awarded for a correct method of recording the data; 1 mark will be awarded for identifying factors to be controlled.

c 1 mark will be awarded for identifying that to avoid bias you should either place quadrats randomly or place them at regular, prefixed points.
1 mark will be awarded for identifying that reliable, reproducible results can be produced by having sufficient repeats, giving a large sample size, etc.

B2 8: Proteins and enzymes

1 They are built from long chains of amino acids, which are folded to give a specific shape.

2 Membrane proteins, hormones, antibodies, enzymes, and structural proteins.

3 They have an active site, which is complementary to the substrate. The substrate fits into the active site, and the reaction occurs.

B2 9: Enzymes and digestion

1 Digestion

2 Amylase breaks down starch into sugars.

3 The enzymes in the stomach work at an acid pH (pH2).

B2 8–9 Levelled questions: Enzymes and digestion

Working to Grade E

1 Amino acids

2 Diagram should be a long chain of beads, and each bead should represent an amino acid.

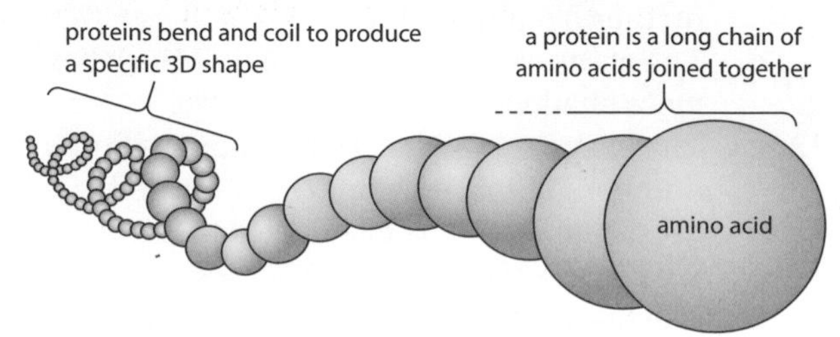

3 A biological catalyst that speeds up the rate of chemical reactions.
4 Proteins
5 The breakdown of large insoluble food molecules into small soluble food molecules.
6 a The liver
b The gall bladder

Working to Grade C

7 The structure allows the specific shape to be formed. The shape is specific to their function

8

Type of protein	Function of protein
Antibodies	Bonds to a pathogen destroying them
Enzymes	**Speed up the rate of chemical reactions**
Membrane proteins	Allows substances into cells through membranes
Hormones	Controls the body's functions

9 Fibres in muscle cells.
10 c, a, d, b, e
11 a To break down large molecules into small ones.
b To build large molecules from small ones.
12 pH
13 The reactions in the organism would be too slow for it to survive.

14

Region of enzyme action in the gut	Enzymes released	Reactions occurring
mouth	amylase	starch → sugars
stomach	**protease**	proteins → amino acids
small intestine	**amylase**	starch → sugars
	protease	**proteins → amino acids**
	lipase	lipids → fatty acids and glycerol (fats and oils)

15 It creates the correct pH for the stomach protease, and kills bacteria entering the gut.
16 It makes enzymes which digest foods in the small intestine.

Working to Grade A*

17 The shape of the active site of the enzyme is complementary to the substrate shape. No other substrate will fit into this active site.
18 a As the temperature increases, the rate of the reaction increases. It reaches an optimum temperature at which the rate is at its highest. Above the optimum temperature, the rate of reaction decreases rapidly, as the enzyme becomes denatured.
b The molecules are moving faster as they have heat energy. They will then bump into each other more, increasing the rate at which they can react together.
c Above the optimum temperature, the increase in temperature begins to damage the shape of the enzyme. The shape of the active site is lost. The enzyme can no longer fit the substrate molecules.
19 To neutralise the acid from the stomach (emulsifies fats).

B2 8–9 Examination questions: Enzymes and digestion

1 a Enzymes speed up the rate of chemical reactions. (1)
b i Stomach (1)
ii 2 marks to be awarded as follows: Enzyme A works best in an acid pH (1) and the stomach has an acid pH (1).
iii The line should continue down in a gradual slope, meeting the axis at about pH 8. (1)

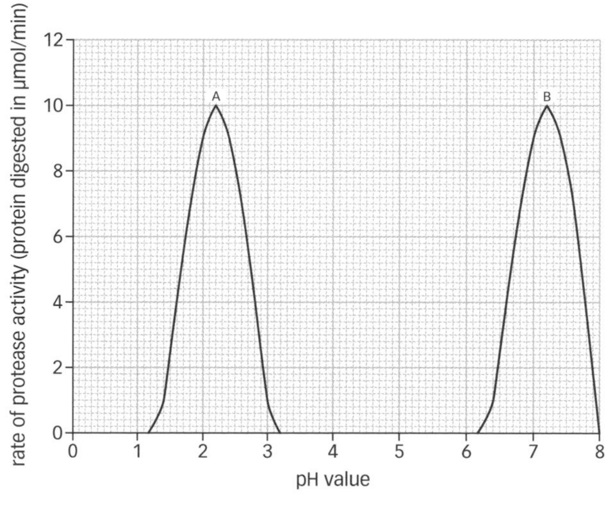

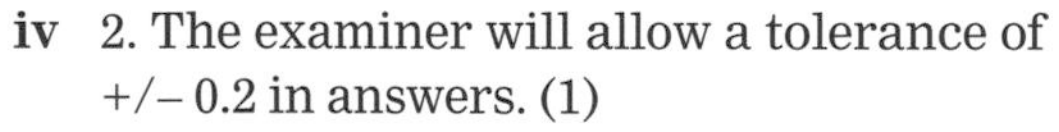

iv 2. The examiner will allow a tolerance of +/– 0.2 in answers. (1)
c 4 marks to be awarded as follows: a higher pH denatures the enzyme (1); this changes the shape of the enzyme (1); this damages the active site (1); so it is no longer complementary to the substrate (1).

B2 10: Applications of enzymes

1 To remove stains from clothes.
2 Proteases, carbohydrases and isomerases.
3 Above 40 °C the enzyme will become denatured and will no longer work. Below 40 °C the rate of reaction will become too slow.

B2 11: Respiration

1 Inside cells
2 Aerobic respiration requires oxygen; anaerobic respiration does not use oxygen.
3 Any two from: heart rate increases, which sends more blood to the tissues; breathing rate and depth increases, which increases oxygen uptake; glycogen is broken down, which releases more glucose to the cells.

B2 12: Cell division

1 Thread-like structures in the nucleus of every cell made of DNA, which contains the genes.
2 In areas of growth, repair, and asexual reproduction.
3 Mitosis produces identical daughter cells; meiosis produces cells with half the number of chromosomes.

B2 13: Stem cells

1 Unspecialised
2 Embryonic cells
3 One issue from:
- Embryonic: destroys an embryo; whether the benefits outweigh the costs; all life is valued.
- Umbilical cells: parents might have a second child for the wrong reasons.

B2 10–13 Levelled questions: Enzymes, respiration, and cell division

Working to Grade E

1 A cleaning agent.
2 Proteases and lipases.
3 a The reactions can be carried out at **lower** temperatures.
  b The reactions can be carried out at **lower** pressures.
  c The reactions will occur at a **higher** rate.
  d The cost of the process will be **lower**.
4 The release of energy from sugars in the cell.
5 Sugars such as glucose.
6 Building larger molecules; muscle contraction; maintaining body temperature.
7 The heart rate increases.
8 a Mitosis
  b Meiosis
  c Mitosis
9 Sperm are made in the testes of males, eggs are made in the ovaries of females.

Working to Grade C

10 Proteases digest proteins, lipases digest fats.
11 They remove stains that non-biological washing powders leave behind. They also remove stains at lower temperatures.
12 Most enzymes work best at 40 °C; above this they are denatured.
13 a Baby food:
  i Substrate is protein.
  ii Enzyme is a protease.
  iii Product: digested proteins/peptides/amino acids.
  Slimming bar:
  i Substrate is glucose.
  ii Enzyme is an isomerase.
  iii Product: fructose.
  b One from: they can't be used at high temperatures; they are expensive to produce.

14

	The reactants	The products
aerobic	glucose and oxygen	carbon dioxide and water
anaerobic	glucose	lactic acid

15 Mitosis
16 a 4
  b 2
  c 4
17 Any two from: embryonic stem cells; bone marrow; umbilical cord blood.
18 An undifferentiated cell (a cell that has not specialised as any one type).
19 Any one from: Parkinson's disease; spinal injuries; organ donation; diabetes.
20 Where cells become specialised to do a particular job.

Working to Grade A*

21 a Baby food: the protein is pre-digested, making it easy for the baby to digest and then absorb the products.
  b Slimming bar: fructose is sweeter than glucose, so less is needed in the slimming bar.
22 They digest cheap starch to produce sugar syrup, which is used in foods.
23 To pump the blood to the muscles quicker. This increases the supply of oxygen and glucose to muscles and takes away carbon dioxide.
24 When the muscles respire anaerobically during a shortage of oxygen, building up lactic acid. Oxygen debt is the amount of oxygen required to get rid of the lactic acid.
25 There is a drop in blood pH due to the build-up of carbon dioxide in the blood. This causes an increase in the breathing rate to increase the uptake of oxygen.

B2 10–13 Examination questions: Enzymes, respiration, and cell division

1 1 mark is awarded for each label, as shown.

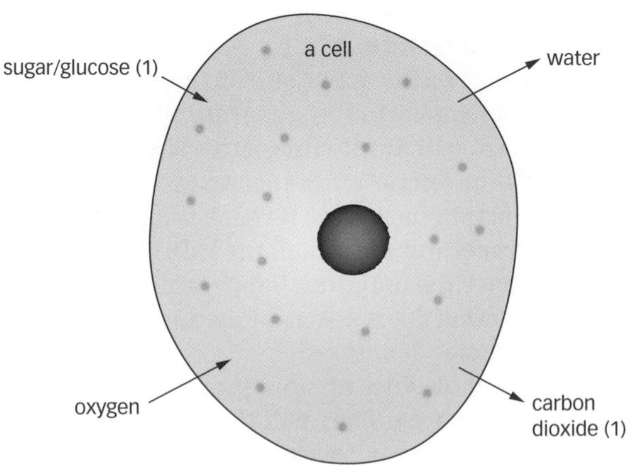

2 Gene, Chromosome, Nucleus, Cell. (3)

3 a Meiosis (1)
b 1 mark is awarded for a diagram completed as shown.

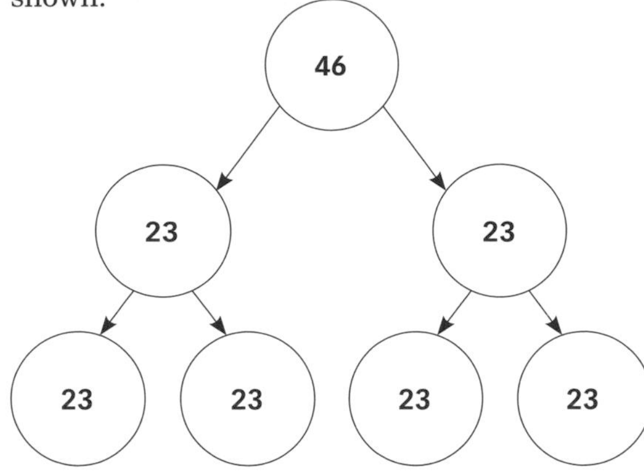

c Gametes (eggs or sperm). (1)

4 a Because they can develop into many other cell types in the body. These can be used in treatment. (1)
b 1 mark for any one from: embryos; umbilical cord blood; bone marrow.
c 1 mark for any one from: destruction of embryos, people having children to act as donors.

B2 14: Inheritance

1 Passing on characteristics from one generation to the next.
2 Gregor Mendel
3 Cross of two heterozygous individuals.

B2 15: Genes and genetic disorders

1 Which of the sex chromosomes we possess: XX = female, XY = male.
2 A recessive allele (c); a person must be homozygous cc to have the disorder.
3 It contains a genetic code. The code determines the order of the amino acids in a protein. This determines how the protein will work.

B2 16: Old and new species

1 Evidence of earlier life forms and how those forms might have changed.
2 Because the environment will change, and species will not be adapted to the new conditions and will die.
3 In the two populations of one species, different characteristics will be favoured in the different conditions. The individuals with these characteristics will survive. Their alleles are passed onto the next generation. Over time, each population becomes so different they cannot interbreed/form new species.

B2 14–16 Levelled questions: Inheritance and evolution

Working to Grade E

1 A section of DNA that controls a characteristic.
2 Mendel was the man who worked out the patterns of inheritance.
3 Pea plants
4 a XY
b XX
c Male
5 Deoxyribonucleic acid
6 A condition or illness caused by a defective gene.
7 The development of additional digits on the hands or feet.
8 The preserved remains of living things from years ago.
9 When a species dies out.

Working to Grade C

10 a Gametes: (T) (T) (t) (t)
b

Gametes	t	t
T	Tt	Tt
T	Tt	Tt

11 a When an allele always controls the development of the characteristic.
b When an allele will only control the development of the characteristic if the dominant allele is not present.
12 a 46
b 23
13 a Chromosomes present: XY × XX
Gametes: (X) (Y) (X) (X)

Gametes	X	X
X	XX	XX
Y	XY	XY

b 50% or 1 in 2.
14 Answer should show the gene located in the same position on the second chromosome.
15 A spiral molecule like a twisted ladder, called a double helix.
16 The sequence of amino acids in a protein.
17 No
18 Identical twins
19 To identify a criminal from tissue evidence at the scene of a crime and to establish family connections such as paternity.
20 Soft bodies tend to rot quickly so there is not enough time for them to fossilise.
21 Some organisms are not well adapted to cope with the new conditions.
22 Speciation is where one species evolves into two new species.
23 It divides the population into two groups; each group then develops or evolves independently of the other.

24 There is very little evidence about the earliest life forms to help scientists explain the process.

25 Hunting

Working to Grade A*

26 a A list of the alleles present as a code.

b A description of the characteristic in words.

c The genotype has identical alleles.

27 Parents: Brown mouse × Brown mouse

Genes present: Bb × Bb

Gametes: (B) (b) (B) (b)

Gametes	B	b
B	BB	Bb
b	Bb	bb

The chance of a white mouse is 1 in 4 or 25%.

28 a Inheritance was controlled by factors; factors are in pairs in adult cells; only one of each pair is in the gamete; offspring have two factors, one from each parent; the outcomes of crosses can be predicted.

b There was a lack of scientific knowledge at the time – scientists had not discovered chromosomes.

c He needed to control pollination, so that he knew which plants were the parents of which offspring.

29 Each new individual is a mix of chromosomes from the parents – half from the mother and half from the father.

30 a Three

b Triplet

31 Each gene has a different sequence of bases. The order of the amino acids in a protein is determined by the sequence of bases in the DNA.

32 a Precious – Pp, Moses – pp

b Parents: Moses × Precious

Phenotype: unaffected × affected

Genotype: pp × Pp

Gametes: (p) (p) (**P**) (p)

Gametes	**P**	p
p	**P**p	pp
p	**P**p	pp

The chances of having a fourth child with polydactyly is 50% or 1 in 2.

33 a When IVF is used to produce the embryos, they can be screened. This means that the cells of the embryo are checked for a specific allele.

b It is discrimination against people with a genetic disorder; embryos with the disorder are discarded, which raises ethical concerns.

34 To avoid contaminating the sample with any other person's DNA.

35 a Bones

b Lack of a complete fossil record.

c Footprints in soft mud that fossilised.

d Global temperature increase that the mammoth was unable to cope with.

36
- Isolation of two groups, one on each side of the river.
- Variation in characteristics develops in each separated population.
- Different characteristics will be favoured in the different populations by natural selection.
- Over time, each group becomes different, until eventually they can no longer interbreed.

B2 14–16 Examination questions: Inheritance and evolution

1 Genes present: RR × rr

Genes in the gametes: R × r

Gametes	R	R
r	Rr	Rr
r	Rr	Rr

1 mark will be awarded for identifying each pair of genes, up to 2 marks.

1 mark will be awarded for identifying each gene in the gametes, up to 2 marks

1 mark will be awarded for correctly completing the table.

2 a Extinction is where a species has no individuals alive today. (1)

b 1 mark will be awarded for each point below, up to a total of 3.
- Restricted or limited diet – they only eat bamboo and they need a lot of it, also it dies back regularly.
- Low birth rate, so not many new individuals produced to breed from, so low numbers in wild and small population.
- Loss of habitat – humans destroy their habitat.

3 1 mark will be awarded for each point below, up to a total of 5.
- The original population of squirrels becomes separated into two groups.
- They become isolated from each other by the canyon.
- Genetic variation occurs.
- Natural selection occurs on both sides of the canyon.
- Different features are favoured on each side of the canyon.
- Over time, the two groups become so different they cannot interbreed.

C2 1: Ionic bonding

1

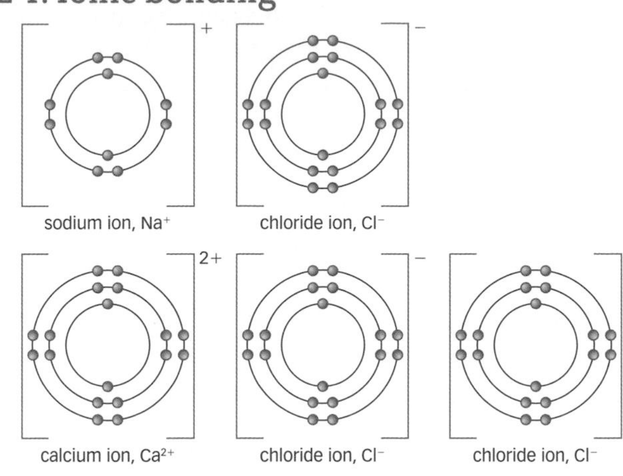

2 sodium + bromine → sodium bromide

3 There are strong electrostatic forces of attraction between oppositely charged sodium and chloride ions. These forces act in all directions. This is ionic bonding.

C2 2: Covalent bonding and metallic bonding

1 Simple molecules consist of a small number of atoms joined together by covalent bonds. Macromolecules are made up of large numbers of atoms joined together by covalent bonds to make a huge network.

2 H—H hydrogen

O=O oxygen

H—Cl hydrogen chloride

O—H (with H below O) water

3 In a giant covalent structure, the atoms are held together by strong covalent bonds to make a huge network. In a giant metallic structure, the electrons in the highest occupied level of the atoms are delocalised. These electrons move through the whole structure of the metal. The positive metal ions are arranged in a regular pattern. The whole structure is held together by strong electrostatic forces of attraction between the positive ions and the moving delocalised electrons.

C2 1–2 Levelled questions: Structure and bonding

Working to Grade E

1 Compounds – **b**, **c**, **f**, and **g**.

2 **a** True
  b When atoms share electrons, they form covalent bonds.
  c True
  d True
  e Elements in group 7 form ions with a charge of –1.

3 ions; positively; ions; negatively; noble gas; zero.

4 **a** Sodium chloride
  b Lithium oxide
  c Potassium iodide
  d Sodium bromide

5 Ionic – middle diagram; simple molecular – bottom diagram; giant covalent – top diagram.

Working to Grade C

6 Formulae that represent compounds are **c** and **e**.

7

Ion	Number of protons	Number of electrons
Li^+	3	2
F^-	9	10
Na^+	11	10
Cl^-	17	18
Mg^{2+}	12	10
Br^-	35	36
Ca^{2+}	20	18

8 **a** $[2.8]^+$
  b $[2.8.8]^-$
  c $[2.8]^{2+}$
  d $[2.8.8]^{2-}$
  e $[2.8.8]^{2+}$
  f $[2.8]^-$

9

Formula of positive ion	Formula of negative ion	Formula of compound
Na^+	Cl^-	NaCl
Mg^{2+}	O^{2-}	MgO
Ca^{2+}	Cl^-	$CaCl_2$
Rb^+	O^{2-}	Rb_2O

10 **a** **b** **c** **d** **e** **f** **g**

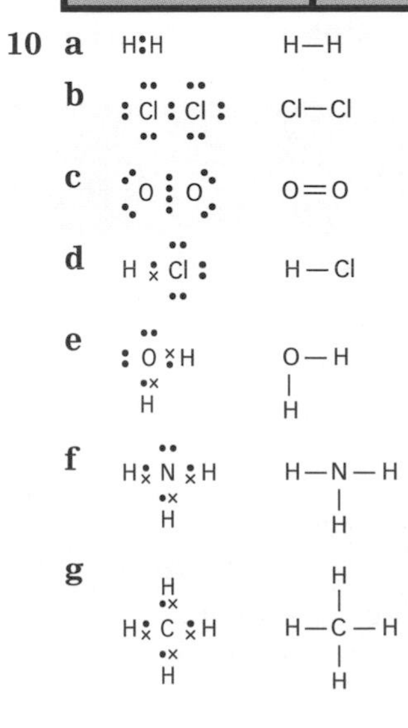

Working to Grade A*

11 In the structure, electrons from the highest energy level of each atom are delocalised. So there is a structure of positive metal ions in a regular pattern. The delocalised electrons move throughout the whole structure. Electrostatic forces of attraction between the positive ions and the delocalised electrons hold the structure together.

C2 1–2 Examination questions: Structure and bonding

1 a i negatively (1), chloride (1).
ii 18 (1)
iii

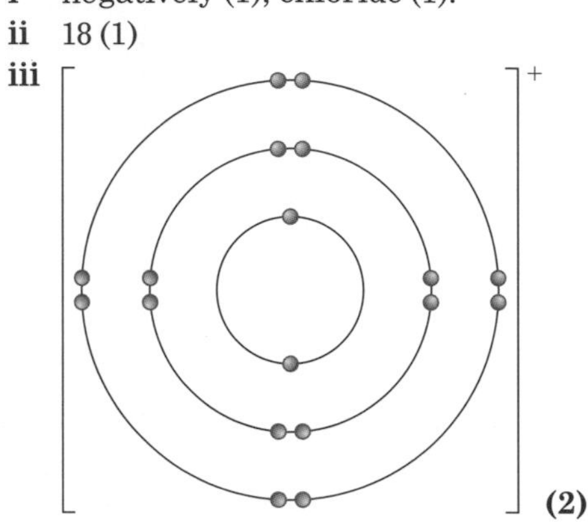

(2)

iv There are strong electrostatic forces of attraction between the oppositely charged ions. (1) These forces act in all directions in the lattice. (1) This type of bonding is called ionic bonding. (1)

b i Metallic bonding (1)
ii Covalent (1)
iii

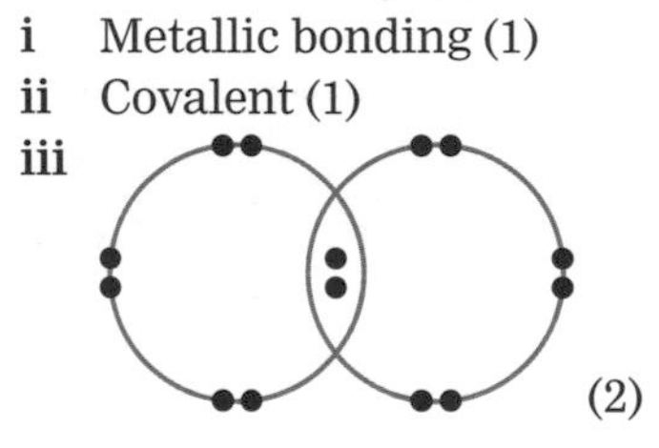

(2)

C2 3: Inside molecules, metals and ionic compounds

1 Substances that consist of simple molecules do not conduct electricity because the molecules do not have an overall electric charge.

2

Type of bonding in substance	Melting and boiling points	Electrical conduction	Other properties
ionic	high	do not conduct when solid; conduct when molten and in solution	–
metallic	high	conduct	easy to bend
simple covalent	low	do not conduct	–

3 Substances that consist of simple molecules have relatively low boiling points because there are only weak intermolecular forces between the molecules. It is these forces that must be overcome when the substances melt or boil, not the much stronger covalent bonds between the atoms in the molecules.

C2 4: The structure of carbon, and nanoscience

1 Diamond is a form of the element carbon. In diamond, each carbon atom forms four covalent bonds with other carbon atoms to make a giant covalent structure.

2 The structure of diamond described in question 1 means that it is very hard. Graphite has a different structure. It is made up of layers. Within each layer, each carbon atom is joined to three other carbon atoms by strong covalent bonds. There are no covalent bonds between the layers, so the layers can slide over each other. This makes graphite soft and slippery.

3 Nanoparticles may be used in new computers, new catalysts, new coatings, highly selective sensors, stronger and lighter construction materials, and new cosmetics.

C2 5: Polymer structures, properties, and uses

1

Type of polymer	Properties	Structure
Thermosoftening	Soften easily when warmed. Can easily be moulded into new shapes.	Consist of individual polymer chains, with weak forces of attraction between the chains.
Thermosetting	Do not melt when heated.	Consist of polymer chains with cross-links between them.

C2 3–5 Levelled questions: Structure, properties, and uses

Working to Grade E

1 low; does not; do not have

2 a The forces between the oppositely changes ions in ionic compounds are strong.
b True
c Ionic compounds conduct electricity when molten and when dissolved in water.
d Ionic compounds do conduct electricity when dissolved in water.
e True

3 atoms, bendy, harder, less

4 a Nitinol
b Dental braces

Working to Grade C

5 a J and N
  b L
  c M
  d M
  e K and M
  f L

6 a Thermosoftening polymers can be recycled because the forces of attraction between their particles are relatively weak, so they melt when heated and can be moulded into new shapes.
  b Thermosetting polymers cannot be recycled because the cross-links between their particles mean that they do not melt on heating. This means they cannot be moulded into new shapes.
  c Low density poly(ethene) and high density poly (ethene) are made under different conditions, and using different catalysts. This gives the two polymers their different properties.

7 The diagram shows that in diamond each carbon atom is joined to four others by strong covalent bonds to make a giant covalent structure. This makes diamond very hard. In graphite, each carbon atom is joined to only three carbon atoms by covalent bonds, to form layers of carbon atoms. The forces of attraction between the layers are very weak, so graphite is soft and slippery.

Working to Grade A*

8 a carbon, hexagonal
  b To deliver drugs to specific targets in the body; as catalysts; as lubricants; to reinforce materials such as graphite tennis racquets.

9 a B
  b G
  c B
  d M
  e B

10 In silicon dioxide, the atoms are held together by strong covalent bonds to form a giant structure. Large amounts of energy are required to overcome these strong covalent bonds, so silicon dioxide has high melting and boiling points. Nitrogen dioxide is made up of molecules. The forces of attraction between the molecules – the intermolecular forces – are weak compared to covalent bonds. It is these bonds which must be overcome when nitrogen dioxide melts or boils, so nitrogen dioxide has relatively low melting and boiling points.

11 Diamond has no delocalised electrons – or other particles with an overall electrical charge – so it cannot conduct electricity. Graphite has delocalised electrons. These particles are free to move so graphite can conduct electricity.

12 There are weak intermolecular forces between the polymer molecules in thermosoftening polymers. It is these forces which must be overcome when thermosoftening polymers melt, so they have low melting points. Thermosetting polymers have strong cross-links between the polymer molecules. These cross-links prevent thermosetting polymers melting when they are heated.

C2 3–5 Examination questions: Structure, properties, and uses

1 a stronger (1); stiffer (1).
  b i 1 (1)
    ii covalent (1)
    iii thermosoftening (1)

2 a 1 – 10 nm (1)
  b Carbon nanotubes are suitable for reinforcing the materials used to make wind turbines because they are very strong when subjected to pulling forces, (1) and very stiff. (1)
  c i Carbon nanotubes may cause cell death, (1) and may cause lung problems. (1)
    ii The tests on human cells were carried out on human cells outside the body and the studies on lung health were carried out on mice and rats (1). This means it is not possible to certain whether similar problems would be caused by nanotubes inside the human body. (1)

3 a 1 mark for each of the following points, up to a maximum of 4:
  - In pure platinum, the atoms are arranged in layers.
  - The layers can slide over each other very easily, making platinum relatively soft.
  - In the alloy, the different sized atoms of rhodium distort the structure of platinum, making it more difficult for them to slide over each other.
  - This makes the platinum-rhodium alloy harder than pure platinum.

  b Rhodium is a good conductor of electricity because it has delocalised electrons to carry the current. (2)

C2 6: Atomic structure

1 Nitrogen: atomic number = 7 and mass number = 14; iron: atomic number = 26 and mass number = 56; bromine: atomic number = 35 and mass number = 80

2 Atoms of the same element can have different numbers of neutrons, and so different mass numbers. Atoms of an element which have different numbers of neutrons are called isotopes.

3 $207 + \{(14 + [16 \times 3]) \times 2\} = 331$

C2 7: Quantitative chemistry

1 $(16 \div 40) \times 100 = 40\%$

2 The maximum theoretical yield may not be obtained because some of the product may be lost when it is separated from the reaction mixture; a reactant might have reacted in an unexpected way; the reaction might be reversible.

3

	Hydrogen	Carbon
Mass of each element, in g	0.4	1.2
A_r from periodic table	1	12
Mass divided by A_r	0.4	0.1
Simplest ratio	4	1
Formula	CH_4	

C2 8: Analysing substances

1 Advantages of instrumental analysis methods include their sensitivity, accuracy, and speed.

2 The sample is heated so that it becomes a mixture of vapours. Then a carrier gas is mixed with the mixture. The carries gas takes the mixture of vapours through a column packed with solid materials. Different substances in the vapour mixture travel through the column at different speeds, and become separated.

C2 6–8 Levelled questions: Atomic structure, analysis, and quantitative chemistry

Working to Grade E

1 Proton = 1; neutron = 1; electron = very small

2 Advantages – **a**, **c**, **d**

3 B, G, D, A, E, C, F

4 **a** True

b The relative formula mass of a substance, in grams, is called one mole of that substance.

c True

Working to Grade C

5 Argon: atomic number = 18 and mass number = 40; manganese: atomic number = 25 and mass number = 55;
zinc: atomic number – 30 and mass number = 65

6 **a** $(12 \times 8) + (1 \times 9) + 14 + (16 \times 2) = 151$ g

b $(12 \times 9) + (1 \times 8) + (16 \times 4) = 180$ g

7 **a** $[39 \div (39 + 12 + 14)] \times 100 = 60\%$

b $\{(7 \times 2) \div [(7 \times 2) + 12 + (16 \times 3)]\} \times 100 = 19\%$

c $\{(14 \times 2) \div [(12 \times 10) + (1 \times 12) + (14 \times 2) + 16]\} \times 100 = 16\%$

8 **a** 4

b D

c A

9 The maximum theoretical yield may not be obtained because some of the product may be lost when it is separated from the reaction mixture; a reactant might have reacted in an unexpected way; the reaction might be reversible.

Working to Grade A*

10 58

11 The relative atomic mass of an element compares the mass of atoms of the element with the ^{12}C isotope. It is an average value for the isotopes of the element.

12 **a**

	Sulfur	Oxygen
Mass of each element, in g	3.2	3.2
A_r from periodic table	32	16
Mass divided by A_r	0.1	0.2
Simplest ratio	1	2
Formula	SO_2	

b

	Carbon	Hydrogen
Mass of each element, in g	2.4	0.4
A_r from periodic table	12	1
Mass divided by A_r	0.2	0.4
Simplest ratio	2	4
Formula	C_2H_4	

c

	Sodium	Nitrogen	Oxygen
Mass of each element, in g	2.3	1.4	4.8
A_r from periodic table	23	14	16
Mass divided by A_r	0.1	0.1	0.3
Simplest ratio	1	1	3
Formula	$NaNO_3$		

13 One mole of calcium carbonate has a mass of 100 g. So 10 g of calcium carbonate is 0.1 mole. The equation shows that one mole of calcium carbonate decomposes to make one mole of carbon dioxide. So 0.1 mole of calcium carbonate makes 0.1 mole of carbon dioxide. The mass of one mole of carbon dioxide is 44 g. So the maximum theoretical yield is 4.4 g.

14 Percentage yield = $(1.0 \div 1.5) \times 100 = 67\%$

C2 6–8 Examination questions: Atomic structure, analysis, and quantitative chemistry

1 **a** **i** 91 (1)

ii 40 (1)

b **i** First atom – number of neutrons = 52; (1) second atom – number of neutrons = 54. (1)

ii isotopes (1)

2 a 1 mark is awarded for your calculation and 1 mark for the correct answer.
$(12 \times 17) + (1 \times 19) + 14 + (16 \times 3) = 285$

b i 369 g (1)

ii 1 mark is awarded for your calculation and 1 mark for the correct answer.
$[(16 \times 5) \div 369] \times 100 = 22\%$

c i Morphine travels more slowly – it has the longer retention time. (1)

ii The mixture is separated. (1)

3 a 1 mark is awarded for your calculation and 1 mark for the correct answer.
$(1.5 \div 2) \times 100 = 75\%$

b Some of the magnesium may have reacted with nitrogen from the air, so producing magnesium nitride as well as magnesium oxide. (1)

How science works: Structures, properties and uses

1 Answer should include advantages and disadvantages such as the following:
Nitinol stents have the advantage of changing shape to match the shape of the blood vessels they are holding open, but stainless steel stents do not. Nitinol also has the advantage of changing shape after being squashed, whereas stainless steel does not. A third advantage of nitinol stents is that blood clots are less likely to form on them than on stainless steel stents. It is possible, but unlikely, that nickel compounds can get into the blood of a person with a nitinol stent. This may increase the risk of cancer. Stainless steel stents contain small amounts of other metals, which may cause blockages in the blood vessel to form near the stent.
Answer should then include a reasoned decision stating which type of stent is better.

2 The hospital might compare the costs of buying and inserting the two different types of stents.

3 a Anita Smith is funding herself, but Rachel Hooper is funded by a company that makes stainless steel stents, so might be influenced by the views of the company.

b Any well-reasoned answer can be given credit, for example: Professor Nadeem Hanif because he has a great deal of experience, and is not funded by an organisation that might bias his views.

C2 9: Rates of reaction and temperature

1 The graph shows that as the temperature increases, the reaction rate also increases.

2 Average rate = $35\,cm^3 \div 2\,min = 17.5\,cm^3/min$

3 The reaction has finished, so no more hydrogen is made.

C2 10: Speeding up reactions: concentration, surface area, and catalysts

1 Factors that affect reaction rate – temperature; concentration of reactants in solution; surface area of solid reactants; pressure for reactions involving a gaseous reaction.

2 The bigger the surface area of a solid reactant, the greater the number of reactant particles that are exposed. The greater the number of particles that are exposed, the greater the frequency of collisions and so the faster the reaction.

3 Catalysts speed up reactions without themselves being used up. They are important in industry because they make reactions fast enough to be profitable.

C2 9–10 Levelled questions: Rates of reaction

Working to Grade E

1 **a** and **c** increase the rate of reaction.

2 The minimum energy the particles must have to react.

3 A catalyst is a substance that speeds up a reaction without itself being used up in the reaction.

4 a The volume of gas increases quickly at first, and then more slowly.

b i Rate = $50\,cm^3 \div 1\,min = 50\,cm^3/min$

ii Rate = $(70 - 50)\,cm^3 \div 1\,min = 20\,cm^3/min$

c

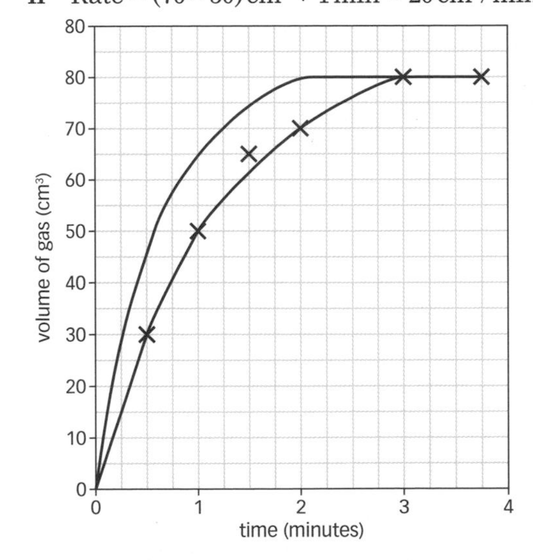

Working to Grade C

5 a Increasing the temperature increases the speed of the reacting particles so that they collide more frequently and more energetically. This increases the rate of reaction.

b Increasing the concentration of reactants in solutions increases the frequency of collisions and so increases the rate of reaction.

6 Using a catalyst helps to make the reaction fast enough to be profitable.

7 a Temperature

b Concentrations of solutions; volumes of solutions; how much the mixture is stirred or agitated.

c Range = 26 s to 400 s

d 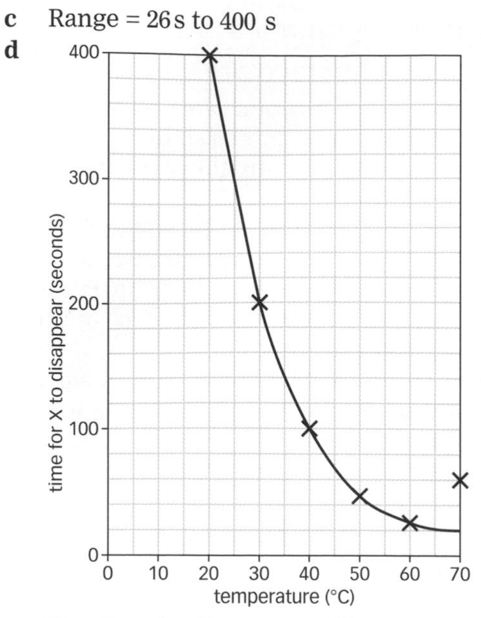

e The time for the cross to disappear at 70 °C is the anomalous result.

f The graph shows that the greater the temperature, the shorter the time for the cross to disappear and so the faster the reaction.

8 a Curve C

b The same amounts of reactants have been used each time.

C2 9–10 Examination questions: Rates of reaction

1 a Two from following list. 1 mark is awarded for each, up to a maximum of 2:
acid concentration, acid volume, amount of magnesium, size of pieces of magnesium.

b i 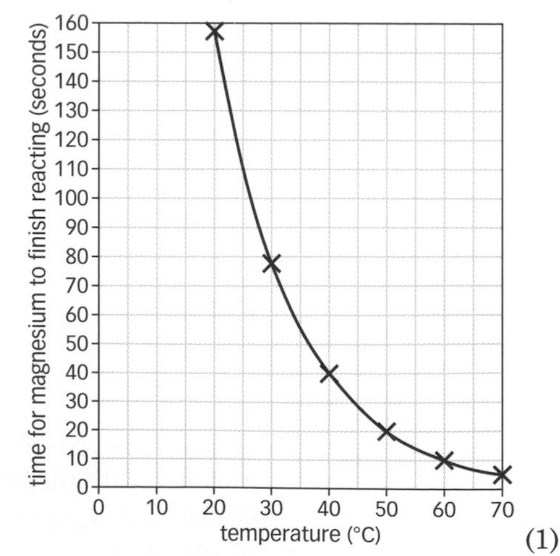

(1)

ii 30 s (1)

iii The graph shows that as the temperature increases, the time for the magnesium ribbon to disappear decreases, showing that the rate has increased. (2)

iv Increasing the temperature increases the speed of the reacting particles so they collide more frequently and more energetically. (1) This increases the rate of reaction. (1)

C2 11: Exothermic and endothermic reactions

1 An exothermic reaction is one that transfers energy to the surroundings.

2 Endothermic reactions can be useful in sports injury packs, to cool damaged muscles.

3 The reaction gives out energy. At first, this energy heats up the reaction mixture. Then heat is transferred from the mixture to the surroundings, and the mixture cools to room temperature.

C2 11: Levelled questions: Exothermic and endothermic reactions

Working to Grade E

1 exothermic, endothermic, always.

2 Endothermic – sports injury packs; exothermic – hand warmers and self-heating cans for coffee.

3 **d** – Thermal decomposition

4 a $(14 + 9 + 10) \div 3 = 11\,°C$

b Anomalous result – test 2 with nitric acid.

c To spot any anomalous data;
To improve the accuracy of the data.

d The volume of the acid and the concentration of the acid.

Working to Grade C

5 Exothermic reactions – **a**, **c**, **d**

6 a A, B, D

b C, E

c C, E

d A, B, D

7 From left to right, the reaction is endothermic and from right to left the reaction is exothermic.

C2 11: Examination questions: Exothermic and endothermic reactions

1 a After the reaction, the temperature of the solution was higher than the temperatures of the reactant solutions before the reaction. This shows that the reaction is exothermic. (1)

b Oxidation and combustion. (1)

c 1 mark for one answer from: hand warmers, self-heating coffee cans.

d i At first, the temperature decreases. (1)

ii The sports injury pack takes in heat energy from the surroundings and returns to the temperature of the surroundings. (1)

C2 12: Acids and bases

1 Bases – copper oxide, zinc oxide, magnesium oxide, sodium hydroxide, potassium hydroxide (or any other metal oxides or hydroxides); alkalis – sodium hydroxide, potassium hydroxide, (or any other soluble metal oxides or hydroxides).

2 Ammonia dissolves in water to make ammonium ions (NH_4^+) and hydroxide ions (OH^-). The hydroxide ions make the solution alkaline.

3 hydrochloric acid + potassium hydroxide → potassium chloride + water

4 HCl (aq) + KOH (aq) → KCl (aq) + H_2O (l)

C2 13: Making salts

1 A salt is a compound that contains metal or ammonium ions. Salts can be made from acids.

2 Add small pieces of zinc to sulfuric acid until there is no more bubbling and a little zinc remains unreacted. Filter to remove the unreacted zinc. Heat the solution over a water bath, until about half its water has evaporated. Leave the solution to stand for a few days to form zinc sulfate crystals.

3 Zinc chloride.

zinc oxide + hydrochloric acid → zinc chloride + water

4 Place some nitric acid in a conical flask. Add a few drops of Universal indicator. Add potassium hydroxide until the solution is just neutral. Add one spatula measure of charcoal powder to remove the colour of the Universal indicator. Filter. Pour the filtrate into an evaporating basin. Heat over a beaker of boiling water until about half the water of the solution has evaporated. Leave the solution to stand for a few days to form potassium nitrate crystals.

nitric acid + potassium hydroxide → potassium nitrate + water

C2 14: Precipitation and insoluble salts

1 A precipitation reaction is one in which two solutions react to form a precipitate.

2 Barium chloride and sodium sulfate (or any other soluble compound of barium and any soluble sulfate).

C2 12–14 Levelled questions: Acids, bases, and salts

Working to Grade E

1 (aq) – dissolved in water; (s) – solid; (l) – liquid; (g) – gas

2
- a chlorides
- b nitrates
- c sulfates

3 alkaline, ammonium, fertilisers

4
- a True
- b Bases are oxides of metals
- c Sulfur dioxide is not a base
- d True
- e True
- f True
- g True
- h True
- i Hydroxide ions make solutions alkaline or hydrogen ions make solutions acidic
- j True
- k A solution with a pH of 6 is acidic

5 C, A, E, B, F, D

Working to Grade C

6

Solution type	pH
Acidic	less than 7
Neutral	7
Alkaline	more than 7

7
- a Magnesium chloride
- b Copper sulfate
- c Potassium nitrate
- d Magnesium sulfate

8 H^+ (aq) + OH^- (aq) → H_2O (l)

9

Solution contains...	acidic, alkaline, or neutral?
An equal number of OH^- and H^+ ions.	neutral
More H^+ ions than OH^- ions.	acidic
More OH^- ions than H^+ ions.	alkaline

10 Add small amounts of magnesium oxide to sulfuric acid until a little solid magnesium oxide remains unreacted. Filter to remove the unreacted magnesium oxide. Heat the solution over a water bath, until about half its water has evaporated. Leave the solution to stand for a few days to form magnesium sulfate crystals.

11 Place some hydrochloric acid in a conical flask. Add a few drops of Universal indicator. Add sodium hydroxide until the solution is just neutral. Add one spatula measure of charcoal powder to remove the colour of the Universal indicator. Filter. Pour the filtrate into an evaporating basin. Heat over a beaker of boiling water until about half the water of the solution has evaporated. Leave the solution to stand for a few days to form sodium chloride crystals.

12 nitrate, potassium, precipitate, lead iodide, potassium nitrate

13
- a Lead iodide
- b Barium sulfate
- c Lead iodide
- d Silver hydroxide
- e Barium sulfate

14 The pairs of solutions given below are examples only; other combinations will also produce the named insoluble salts.
- a Lead nitrate and sodium chloride
- b Calcium nitrate and sodium sulfate
- c Lead nitrate and sodium sulfate
- d Barium chloride and sodium sulfate

C2 12–14 Examination questions: Acids, bases, and salts

1 a Hydrochloric acid (1)
b To remove unreacted zinc metal. (1)
c Crystallisation (1)
d Two from following list. 1 mark awarded for each up to a maximum of 2:
do not touch the product; wash hands after the practical; wear eye protection; work in a well-ventilated laboratory; follow instructions for the disposal of chemicals.
i Copper oxide, magnesium oxide, sodium hydroxide (1)
ii Sodium hydroxide (1)
b Name: hydroxide; (1) formula: OH^- (1)
3 a 1 mark for each of: lead nitrate and potassium iodide solutions
b Mix the two solutions. Filter. The solid lead iodide remains in the filter paper. (2)
c $Pb^{2+}(aq) + 2\,I^-(aq) \rightarrow PbI_2\,(s)$
1 mark awarded for each correct state symbol.

C2 15: Electrolysis 1

1 Positive electrode – chlorine; negative electrode – copper.
2 Positive electrode – oxygen; negative electrode – hydrogen.
3 Positive electrode: $Cl^- \rightarrow Cl + e^-$
then $Cl + Cl \rightarrow Cl_2$
Negative electrode: $Cu^{2+} + 2e^- \rightarrow Cu$

C2 16: Electrolysis 2

1 To protect a metal object from corrosion by coating it with an unreactive metal that does not easily corrode; to make an object look attractive.
2 Aluminium and carbon dioxide.
3 Positive electrode – chlorine gas; negative electrode – hydrogen gas.

C2 15–16 Levelled questions: Electrolysis

Working to Grade E

1 **a** and **b**
2 melted/dissolved; melted/dissolved; ions; solution
3 electrolyte – a liquid or solution that is broken down when electricity passes through it;
electrolysis – the process by which electricity breaks down a liquid or solution;
electroplating – covering an object with a layer of metal in an electrolysis cell;
electrodes – piece of metal or graphite through which electricity enters or leaves an electrolysis cell
4 bromide; positively; positive; negative

Working to Grade C

5 a At the negative electrode, positively charged ions gain electrons.
b True
c If an ion gains electrons, the ion is reduced.
d True

6

Solution	Positive electrode	Negative electrode
copper chloride	chlorine	copper
potassium bromide	bromine	hydrogen
silver nitrate	oxygen	silver
magnesium nitrate	oxygen	hydrogen
copper carbonate	oxygen	copper
sodium sulfate	oxygen	hydrogen

7 Positive electrode – chlorine gas, used for making bleach, plastics, and sterilising water;
negative electrode – hydrogen gas, used for making margarine and ammonia;
solution of sodium hydroxide formed in electrolyte vessel, used for making soap.
8
- Positive electrode **4**
- Positive ions gain electrons at this electrode **1**
- Negative electrode **1**
- Electrolyte of liquid aluminium and cryolite **2**
- Liquid aluminium forms at this electrode **3**
- Oxide ions are oxidised at this electrode **4**
- Carbon dioxide gas forms here **4**
- This electrode is made of carbon **4**
- Reduction happens at this electrode **1**
- Negative ions are attracted to this electrode **4**

Working to Grade A*

9 a Negative electrode: $Pb^{2+} + 2e^- \rightarrow Pb$
Positive electrode: $Br^- \rightarrow Br + e^-$
then $Br + Br \rightarrow Br_2$
b Negative electrode: $Cu^{2+} + 2e^- \rightarrow Cu$
Positive electrode: $Cl^- \rightarrow Cl + e^-$
then $Cl + Cl \rightarrow Cl_2$
c Negative electrode: $Al^{3+} + 3e^- \rightarrow Al$
Positive electrode: $O^{2-} \rightarrow O + 2e^-$
then $O + O \rightarrow O_2$

C2 15–16 Examination questions: Electrolysis

1 a The rhodium is harder, so protects the silver from scratches; (1) the rhodium does not react with gases from the air, so it protects the silver from corrosion. (1)
b i So that positive rhodium ions are attracted towards it, and then gain electrons to form rhodium metal on the surface of the ring. (1)
ii Reduction (1)
iii $Rh^{3+} + 3e^- \rightarrow Rh$ (1)

How Science Works: Rates, energy, salts, and electrolysis

1 a Dependent variable – time to collect 50 cm^3 of hydrogen gas; independent variable – temperature; control variables – (three from) mass of magnesium, surface area of magnesium, concentration of acid, volume of acid, stirring.

b 48

c Graph with temperature on x-axis and time to collect 50 cm^3 hydrogen gas on y-axis. A smooth curve should be drawn.

d The results do support the hypothesis as the rate of reaction increases with increasing temperature.

2 a A new catalyst does not have to be bought each time – this reduces costs.

b The catalyst does not have to be disposed of; smaller amounts of catalyst are required, so reducing the impact on the environment of producing the catalyst material.

P2 1: Speed and velocity

1 5 m/s

2 a B

b C

3 3.25 m/s

P2 2: Acceleration

1 1.5 m/s^2

2 a 0.75 m/s^2

b The line is steeper.

3 48 m

P2 3: Force and acceleration

1 600 N upwards.

2 9 N to the left.

3 3.33 m/s^2

P2 4: Forces and braking

1 75 m

2 wet roads, poor car maintenance, inebriation, lack of attention

3 The resistive forces increase from zero at the start to 800 N at the top speed. This is because air resistance increases with increasing speed.

P2 5: Terminal velocity

1 241 N

2 On the edge the resultant force is zero as the reaction force from the ground balances out your downwards force from gravity. Over the edge, the only force is gravity, so you accelerate down.

3 From A to B the force of gravity accelerates the skydiver downwards. As the speed increases, so does the upwards air resistance, reducing the resultant force and the consequent acceleration. By point C the air resistance is big enough to cancel the force of gravity, so the speed doesn't change and the skydiver has reached terminal velocity. At D the parachute is opened, suddenly increasing the air resistance. The resultant force is now upwards, so the skydiver slows down rapidly until at E the air resistance is once more equal and opposite to the force of gravity.

P2 1–5 Levelled questions: Force and motion

Working to Grade E

1 a Speed increases.

b Speed decreases.

c Speed stays the same.

2 2700 N

3 50 m/s

4 zero, accelerates, steady, decelerates, steady

5 Any two from: poor car maintenance, wet road, inattention, inebriation

6 weight, air resistance, balanced, velocity

Working to Grade C

7 5 m/s² to the left

8 20 N downwards, balanced by the upwards reaction from the horizontal surface so that the resultant force in the vertical direction is zero.

9 –0.33 m/s²

10 27.2 m/s, so slower than speed limit.

11 Speed is increasing with time.

12

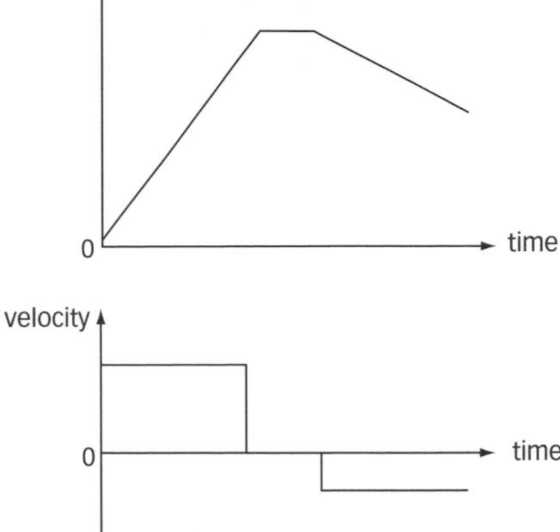

velocity
0
time

13 Braking distance: distance moved by a car while the brakes have been fully applied.
Stopping distance: total distance moved by the car as it is brought to a halt by the driver.
Thinking distance: distance moved by the car in the time between the driver deciding to stop and applying the brakes.

Working to Grade A*

14 +6 m/s, 0 m/s and –2 m/s

15 675 m

P2 1–5 Examination questions: Force and motion

1 One mark for each correct link.

distance (m)
0
0
time (s)
constant acceleration
constant deceleration
constant zero velocity
constant non-zero velocity

2 a The car has zero vertical acceleration (1) so the weight and reaction force from the ground must be equal and opposite to add up to zero (1).

b 1000 – 750 = 250 N (1) to the left (1).

c Mass = 8000 N/10 N/kg = 800 kg (1).
acceleration = force/mass (1) = 250/800 = 0.31 m/s² backwards (1).

d Velocity decreases (1) because the resultant force is against the direction of motion (1).

3 a 44 m (1)

b Time = distance/speed (1) = 6/10 = 0.6 s (1)

c i The brakes provide a steady decelerating force (1), so as the speed increases it takes longer to reduce it to zero (1). If the car is moving for longer, then it must move further; the higher speed of the car will also make it move further in the same time (1).

ii EITHER If the road is wet (1) then there is less friction between the road and the tyres (1) OR if the brakes are faulty (1) then the force slowing the car down is smaller (1) OR if you are going uphill (1) gravity helps to slow down the car (1).

P2 6: Work and energy

1 4800 J

2 240 000 J

3 333 N/m

P2 7: Energy transfer and power

1 120 W

2 4320 J

3 20 m/s

P2 8: Momentum and car safety

1 364 kg m/s

2 Increases the time needed to stop the car, so reducing the acceleration and force applied to the driver.

3 3 m/s

P2 6–8 Levelled questions: Work and energy

Working to Grade E

1 15 N

2 J, kg m/s, W and J (in order)

3 kinetic, heat, kinetic, gravitational

4 5000 W

5 Loses 1 000 000 J

6 Crumple zones, seat belts and airbags.

7 10 kg m/s

Working to Grade C

8 The seatbelt provides the force to slow down the driver over all the time that the car is slowing down. Without the seatbelt the driver would continue moving forwards until they hit the dashboard – slowing down in a shorter time by a much larger force.

9 5000 W
10 41 250 W
11 The total momentum of a set of objects remains constant, provided that they only interact with each other.
12 The extension of a spring is proportional to the force on it, providing that the spring has not been extended too far.
13 800 N/m
14 acceleration, force, momentum and velocity
15 The kinetic energy of the car is stored in a battery as the car slows down. This energy can be used to accelerate the car when it starts moving again, instead of transferring energy from fuel. So less fuel is used up by a car which slows down and accelerates as it goes along.

Working to Grade A*

16 0.6 m/s. Trolleys only interact with each other.
17 If all the GPE transfers to KE during the fall, then speed = 20 m/s.

P2 6–8 Examination questions: Work and energy

1 a

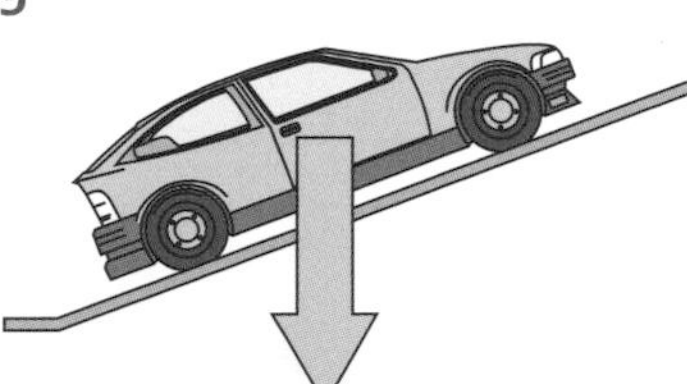

arrow pointing down from centre of car (1)

b $W = mg$ (1) = 1200 × 10 = 12 000 N (1)

c Work = force × distance (1) = 12 000 × 5 = 60 000 J (1)

d GPE (1)

2 a $KE = \frac{1}{2}mv^2$ (1) = 0.5 × 900 × $(15)^2$ = 101 250 J (1)

b $P = \frac{E}{t}$ (1) = 101 250/5 = 20 250 W (1)

c When the car has a crash, a force must act on the driver to slow them down during the crash. If that force is too large, then it will damage the driver. The force can be reduced by reducing the acceleration required. This can be done by increasing the time taken to reduce the car from its initial speed to zero. The crumple zone between the front of the car and the driver does this by collapsing slowly, so that the car can keep moving for a short while after it has hit the obstacle instead of stopping instantly. (6 marks for a clear ordered answer with all steps clearly explained, no errors of physics and good use of spelling and grammar).

3 a $p = m \times v$ (1) = 0.025 × 400 = 10 kg m/s (1)

b i Total momentum (1)

ii $v = \frac{p}{m}$ (1) = 10/2.025 = 4.94 m/s (1)

How science works: Energy and distance

1 Independent variables are mass and initial speed, dependent variable is skid distance.
2 The surface that the cylinders skid on.
3 He needs to calculate kinetic energy from the mass and speed using $KE = \frac{1}{2}mv^2$.
4 A scatter-graph of kinetic energy against skid distance.
5 He won't be able to repeat that trial exactly, so he just has to leave the result on the graph and get lots more results to put on it.
6 Yes, provided that his cylinders are made of rubber and the surface they slide on is made of tarmac.

P2 9: Static electricity

1 Electrons can move easily through a conductor but not an insulator.
2 The rod is negative, the cloth is positive.
3 Electrons transfer between the balloon and the hair. So all of the hairs have the same charge, repelling each other and standing on end. The hair and balloon have opposite charges, so they attract each other.

P2 10: Current electricity

1 3 A

2

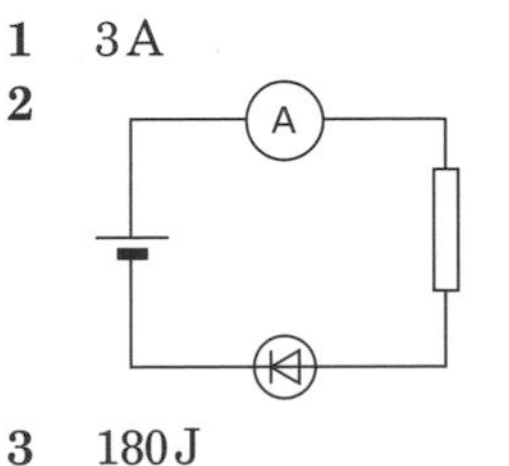

3 180 J

P2 11: Resistance

1 1.4 V

2

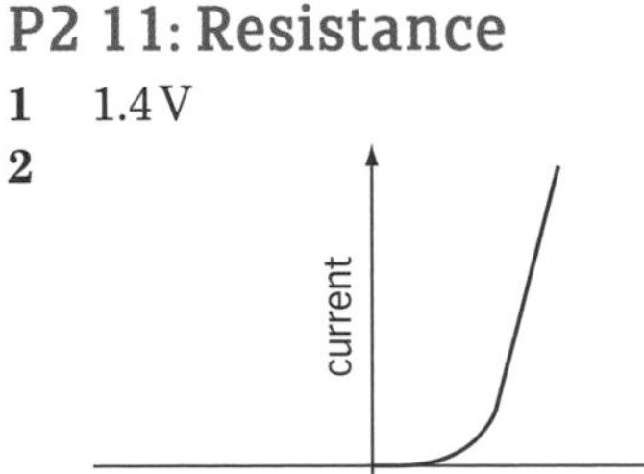

Small resistance in one direction of current, but a very large resistance for current in the other way.

3

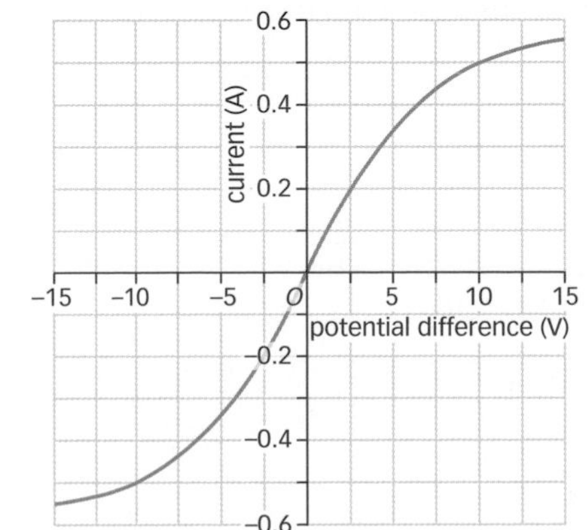

As electrons flowing through the wire hit the ions fixed in position, they give them vibrational energy, heating up the wire. As the temperature of the wire increases, the ions vibrate further from their position, making it more difficult for the electrons to get past them, increasing the resistance of the wire.

P2 12: Circuits in series and parallel

1

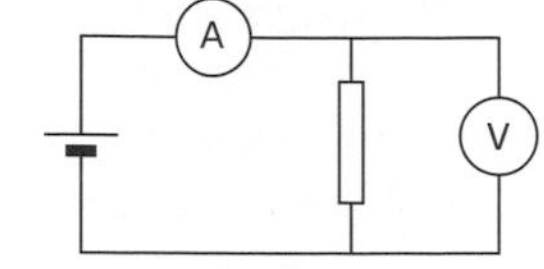

2 Total resistance = 9 Ω, 0.167 A for both, 1.0 V for 6 Ω and 0.5 V for 3 Ω.
3 For 6 Ω resistor, current = 0.25 A and for 3 Ω resistor, current = 0.5 A, so total resistance 2.0 Ω

P2 13: Mains electricity

1 The central cores are made from a bundle of copper threads to make it flexible and a good conductor. Each core is surrounded by a different coloured plastic to provide a flexible insulator which can be identified easily.
2 d.c. supplies make the current flow in one direction only, a.c. supplies make the current change direction all the time.
3 If the live wire touches the outer metal of a device, then there is a large current which escapes in the earth wire connected to the outer metal. This current heats up the fuse in the live part of the plug, causing it to melt and isolate the live part of the plug from the rest of the device.

P2 14: Electrical power

1 The 13 A cable has a larger diameter so that it has a smaller resistance and doesn't get too warm in use.
2 13 A
3 600 C

P2 9–14 Levelled questions: Electricity

Working to Grade E

1 The hair becomes negatively charged.
2 A flow of electrons along a wire.
3

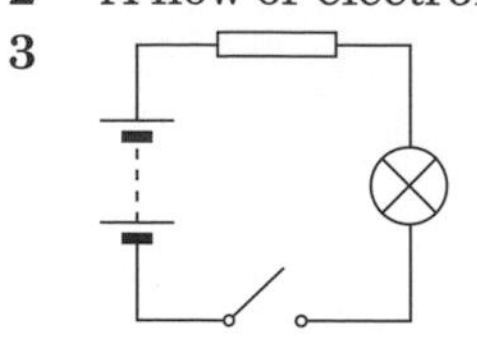

4 It shows an ammeter in series with a variable resistor and LED and cell.
5 180 V
6

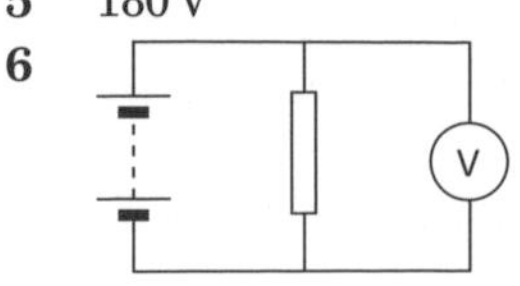

7 Light dependant resistor.
8 blue – neutral, brown – live, yellow and green – earth
9 690 W

Working to Grade C

10 Hair is negative, balloon is positive.
11 2 A
12

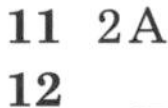

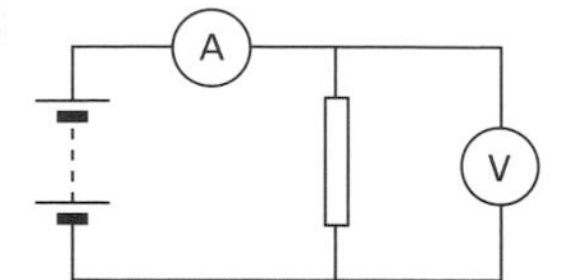

13 A is a diode, B is a lamp and C is a resistor.
14 8 A in the 6 Ω resistor and 12 A in the battery.
15 To switch off the supply automatically when the current is large enough e.g. when there is a fault. The current heats up the wire in the fuse enough to melt it and make a gap in the circuit.

Working to Grade A*

16 75 V and 25 Hz.

P2 9–14 Examination questions: Electricity

1 a A is battery (1), B is voltmeter (1) and C is switch (1).
b $V = I \times R$ (1) = 0.4 × 5 = 2.0 V (1)
c

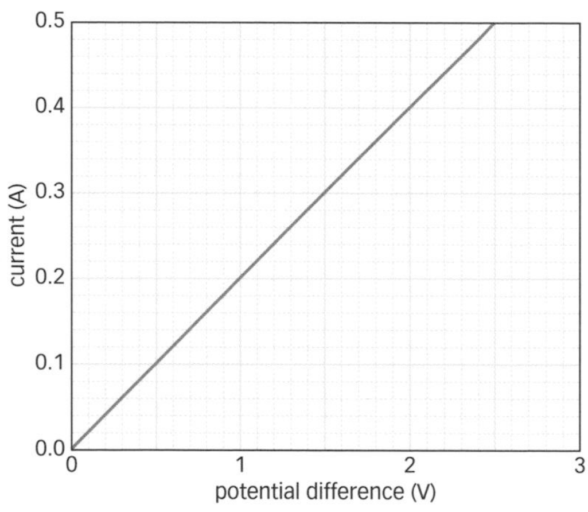

Straight line through the origin (1) passing through 0.40 A, 2.0 V (1).

2 a makes the kettle safe – green and yellow (1)
carries current at 230 V – brown (1)
carries low-voltage current – blue (1)
b $P = V \times I$ (1) = 230 (1) × 7 = 1610 W (1)
c 13 A (1)
d It acts as an automatic switch (1) which disconnects the supply (1) whenever the current in the plug exceeds a safe value (1).
e i Electrons (1) flow easily through a conductor but not through an insulator (1).
ii

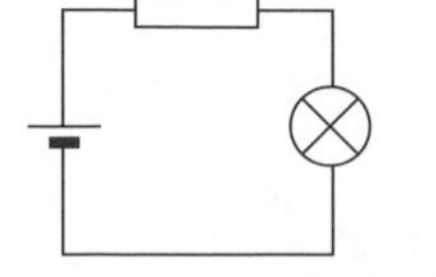

Correct symbols for components (1 mark each) drawn in a loop in any order (1).

P2 15: Atomic structure

1 Electron, proton, neutron.
2 38 electrons, 38 protons and 52 neutrons
3 5

P2 16: Radioactive decay

1 Wear protective clothing, don't handle directly, keep the exposure time short.
2 Not alpha because it goes through paper. Not gamma because it is stopped by aluminium, so must be beta.
3 One hour.

P2 17: Using radioactivity

1 Uranium and plutonium.
2 Only gamma rays escape from the body, carrying information about its interior. A short half-life ensures that the patient is not radioactive for long, reducing the danger from this exposure to people around them.
3 The number of beta particles getting through the paper decreases as the paper thickness increases. All gamma rays would go straight through and none of the alpha particles would get through. A long half-life means that the number of betas getting through a given thickness of paper hardly changes with time.

P2 18: Star lives

1 nebula, protostar, red giant star, white dwarf
2 Gas in the nebula is pulled in by gravity to make a protostar. The work done by gravity heats the gas until it is hot enough for hydrogen to fuse into helium, making a stable star.
3 The main sequence is where hydrogen is fusing to make helium in the core. The energy released by fusion at the centre moves up through the rest of the star, counteracting the push of gravity that is trying to squash it smaller.

P2 15–18 Levelled questions: Radioactivity

Working to Grade E

1 Electron (0, –1)
Proton (1, 1)
Neutron (1, 0)
2 Neutron and proton.
3 Food, soil, space.
4 Alpha
5 Wear protective clothing, don't handle directly, keep the exposure time short.
6 Gamma ray with a half-life of a few hours or days.
7 uranium, neutron, fission
8 protostar, main sequence, super red giant, supernova, black hole

Working to Grade C

9 10 electrons, 10 protons and 12 neutrons.
10 Same number of protons and electrons but different numbers of neutrons.
11 250
12 Their radiation ionises cells, which can kill the cell or change its DNA so that it turns into cancer.
13 Wear protective clothing so that you don't ingest particles of the source. Don't handle directly but use tongs to keep your distance. Keep the exposure time short to reduce the changes of damaging your cells.
14 Beta and gamma.
15 Only gamma rays escape from the body, carrying information about its interior. A short half-life ensures that the patient is not radioactive for long, reducing the danger from this exposure to people around them.
16 Gas in a nebula is pulled in by gravity to make a protostar. The work done by gravity heats the gas until it is hot enough for hydrogen to fuse into helium, making a stable star. Once all the hydrogen in the core has been used the star swells to become a red giant and helium is fused instead. When the helium has been used the star starts to cool, collapsing to form a white dwarf. Eventually, as it gets cold enough, the dwarf stops glowing altogether.

Working to Grade A*

17 2 hours
18 $^{237}_{93}Np$
19 One of the neutrons becomes a proton and an electron. The electron carries away energy from the nucleus, so the change cannot be reversed.

P2 15–18 Examination questions: Radioactivity

1 a Any two of the following, for 1 mark each: food, soil, air, X-rays, cosmic rays, nuclear power stations, nuclear bomb tests.
b It damages cells by ionising them (1).
c Allowed dose from work = 20 – 2 = 18 mSv (1) working hours = 18/0.02 (1) = 900 hours (1).
2 a $^{6}_{3}Li$ (1 mark for each correct number)
b

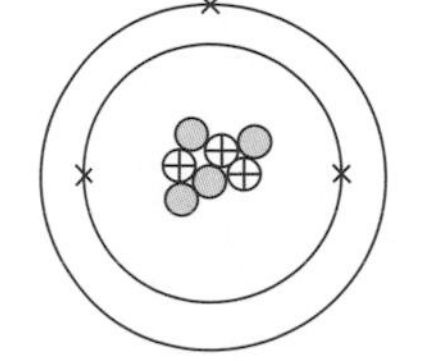

(two crosses in inner circle, one on the other for 1 mark)
(three crossed circles at the centre for 1 mark)
(four shaded circles at the centre for 1 mark)
c fission (1)
d The alpha particles ionise (1) the air in the detector allowing it to conduct electricity (1). A long-half life means that the detector keeps working for a long time (1).

e Wear protective clothing (1) to stop accidental ingestion (1).

f Beta – High speed electron (1), Alpha – Helium nucleus (1), Gamma – Electromagnetic waves (1)

3 a

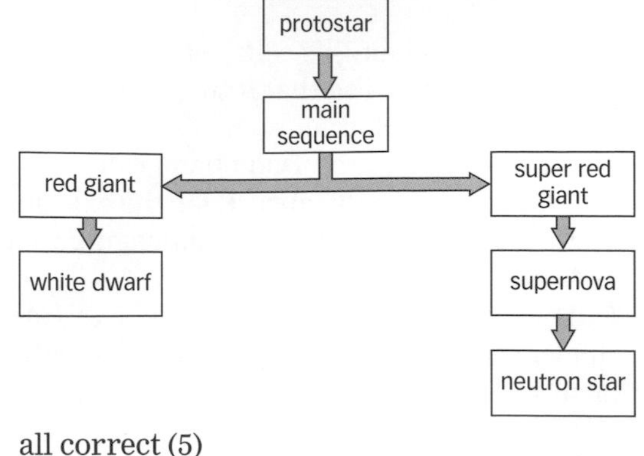

all correct (5)
one mistake (4)
two mistakes (3)
three mistakes (2)
four mistakes (1)

b Hydrogen nuclei (1) are fused together (1) to make helium nuclei (1).

c Some elements with atomic number below iron were made by fusion within stars (1). The rest were made by fusion when large stars exploded as supernovae (1). This happens when a star runs out of fuel for fusion and collapses (1).

How science works: Radioactivity

1 The independent variable is the sheet thickness. The dependent variable is the gamma ray count.

2 The distance between source and detector, the source itself.

3 Calculate the average of three trials for each sheet thickness.

4 Calculate the thickness of each sheet times the average rate of gamma rays getting through. If the answer is the same every time, then the hypothesis is true.

5 She should count the gamma rays for each thickness for a longer time.

Appendices

Periodic table

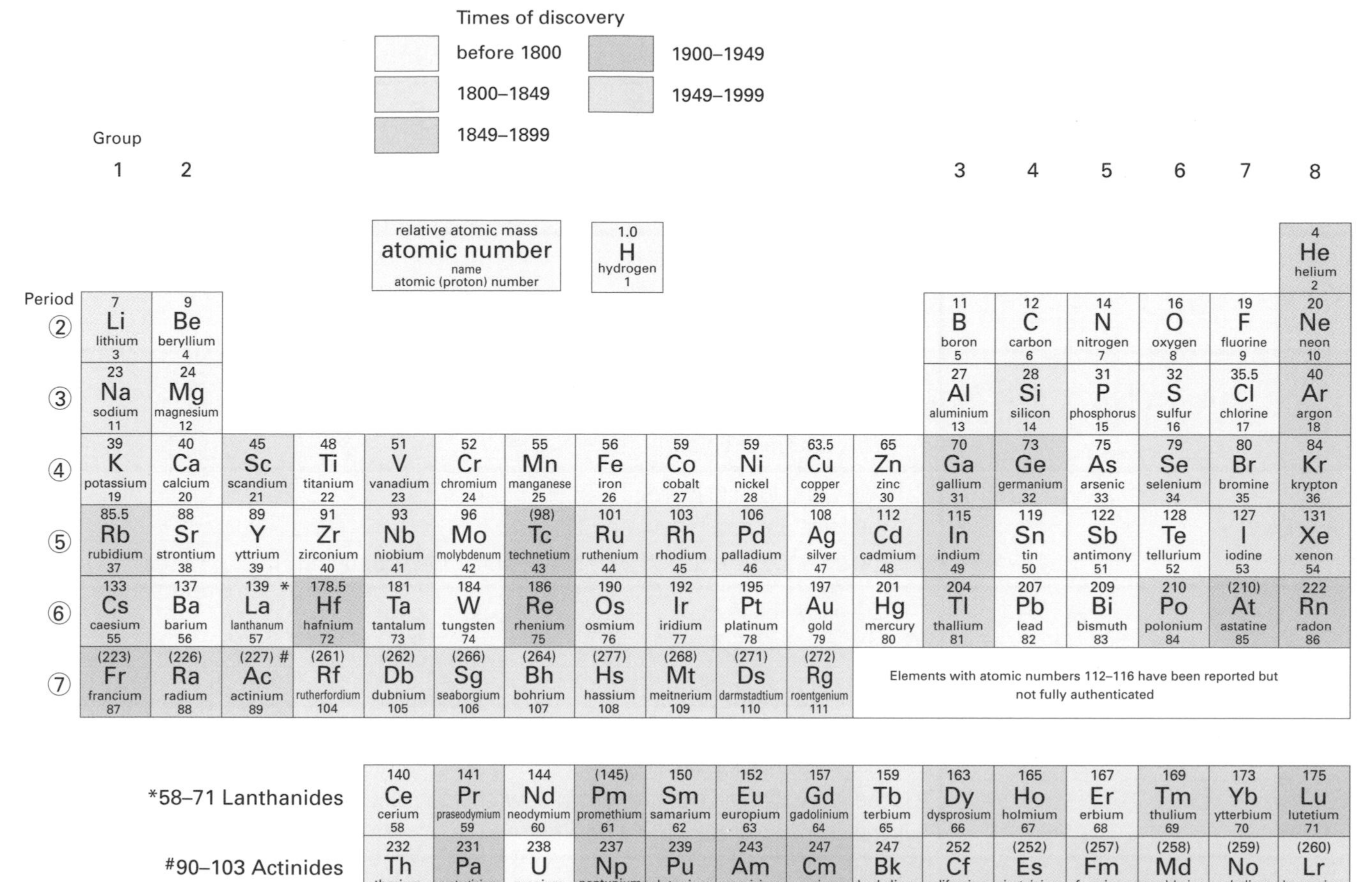

Period	1	2											3	4	5	6	7	8
								1.0 H hydrogen 1										4 He helium 2
②	7 Li lithium 3	9 Be beryllium 4											11 B boron 5	12 C carbon 6	14 N nitrogen 7	16 O oxygen 8	19 F fluorine 9	20 Ne neon 10
③	23 Na sodium 11	24 Mg magnesium 12											27 Al aluminium 13	28 Si silicon 14	31 P phosphorus 15	32 S sulfur 16	35.5 Cl chlorine 17	40 Ar argon 18
④	39 K potassium 19	40 Ca calcium 20	45 Sc scandium 21	48 Ti titanium 22	51 V vanadium 23	52 Cr chromium 24	55 Mn manganese 25	56 Fe iron 26	59 Co cobalt 27	59 Ni nickel 28	63.5 Cu copper 29	65 Zn zinc 30	70 Ga gallium 31	73 Ge germanium 32	75 As arsenic 33	79 Se selenium 34	80 Br bromine 35	84 Kr krypton 36
⑤	85.5 Rb rubidium 37	88 Sr strontium 38	89 Y yttrium 39	91 Zr zirconium 40	93 Nb niobium 41	96 Mo molybdenum 42	(98) Tc technetium 43	101 Ru ruthenium 44	103 Rh rhodium 45	106 Pd palladium 46	108 Ag silver 47	112 Cd cadmium 48	115 In indium 49	119 Sn tin 50	122 Sb antimony 51	128 Te tellurium 52	127 I iodine 53	131 Xe xenon 54
⑥	133 Cs caesium 55	137 Ba barium 56	139 * La lanthanum 57	178.5 Hf hafnium 72	181 Ta tantalum 73	184 W tungsten 74	186 Re rhenium 75	190 Os osmium 76	192 Ir iridium 77	195 Pt platinum 78	197 Au gold 79	201 Hg mercury 80	204 Tl thallium 81	207 Pb lead 82	209 Bi bismuth 83	210 Po polonium 84	(210) At astatine 85	222 Rn radon 86
⑦	(223) Fr francium 87	(226) Ra radium 88	(227) # Ac actinium 89	(261) Rf rutherfordium 104	(262) Db dubnium 105	(266) Sg seaborgium 106	(264) Bh bohrium 107	(277) Hs hassium 108	(268) Mt meitnerium 109	(271) Ds darmstadtium 110	(272) Rg roentgenium 111	Elements with atomic numbers 112–116 have been reported but not fully authenticated						

*58–71 Lanthanides	140 Ce cerium 58	141 Pr praseodymium 59	144 Nd neodymium 60	(145) Pm promethium 61	150 Sm samarium 62	152 Eu europium 63	157 Gd gadolinium 64	159 Tb terbium 65	163 Dy dysprosium 66	165 Ho holmium 67	167 Er erbium 68	169 Tm thulium 69	173 Yb ytterbium 70	175 Lu lutetium 71
#90–103 Actinides	232 Th thorium 90	231 Pa protactinium 91	238 U uranium 92	237 Np neptunium 93	239 Pu plutonium 94	243 Am americium 95	247 Cm curium 96	247 Bk berkelium 97	252 Cf californium 98	(252) Es einsteinium 99	(257) Fm fermium 100	(258) Md mendelevium 101	(259) No nobelium 102	(260) Lr lawrencium 103

Reactivity series of metals

Potassium	most reactive
Sodium	↑
Calcium	
Magnesium	
Aluminium	
Carbon	
Zinc	
Iron	
Tin	
Lead	
Hydrogen	
Copper	
Silver	
Gold	↓
Platinum	least reactive

(elements in italics, though non-metals, have been included for comparison)

Formula of some common ions

Name	Formula	Name	Formula
Hydrogen	H^+	Chloride	Cl^-
Sodium	Na^+	Bromide	Br^-
Silver	Ag^+	Fluoride	F^-
Potassium	K^+	Iodide	I^-
Lithium	Li^+	Hydroxide	OH^-
Ammonium	NH_4^+	Nitrate	NO_3^-
Barium	Ba^{2+}	Oxide	O^{2-}
Calcium	Ca^{2+}	Sulfide	S^{2-}
Copper(II)	Cu^{2+}	Sulfate	SO_4^{2-}
Magnesium	Mg^{2+}	Carbonate	CO_3^{2-}
Zinc	Zn^{2+}		
Lead	Pb^{2+}		
Iron(II)	Fe^{2+}		
Iron(III)	Fe^{3+}		
Aluminium	Al^{3+}		

Equations	
$a = \frac{F}{m}$ or $F = M \times a$	F is the resultant force in newtons, N m is the mass in kilograms, kg a is the acceleration in metres per second squared, m/s^2
$a = \frac{v - u}{t}$	a is the acceleration in metres per second squared, m/s^2 v is the final velocity in metres per second, m/s u is the initial velocity in metres per second, m/s t is the time taken in seconds, s
$W = m \times g$	W is the weight in newtons, N m is the mass in kilograms, kg g is the gravitational field strength in newtons per kilogram, N/kg
$F = k \times e$	F is the force in newtons, N k is the spring constant in newtons per metre, N/m e is the extension in metres, m
$W = F \times d$	W is the work done in joules, J F is the force applied in newtons, N d is the distance moved in the direction of the force in metres, m
$P = \frac{E}{t}$	P is the power in watts, W E is the energy transferred in joules, J t is the time taken in seconds, s
$E_p = m \times g \times h$	E_p is the change in gravitational potential energy in joules, J m is the mass in kilograms, kg g is the gravitational field strength in newtons per kilogram, N/kg h is the change in height in metres, m
$E_K = \frac{1}{2} \times m \times v^2$	E_K is the kinetic energy in joules, J m is the mass in kilograms, kg v is the speed in metres per second, m/s
$p = m \times v$	p is the momentum in kilograms metres per second, kg m/s m is the mass in kilograms, kg v is the velocity in metres per second, m/s
$I = \frac{Q}{t}$	I is the current in amperes (amps), A Q is the charge in coulombs, C t is the time in seconds, s
$V = \frac{W}{Q}$	V is the potential difference in volts, V W is the work done in joules, J Q is the charge in coulombs, C
$V = I \times R$	V is the potential difference in volts, V I is the current in amperes (amps), A R is the resistance in ohms, Ω
$p = \frac{E}{t}$	P is power in watts, W E is the energy in joules, J t is the time in seconds, s
$P = I \times V$	P is power in watts, W I is the current in amperes (amps), A V is the potential difference in volts, V
$E = V \times Q$	E is the energy in joules, J V is the potential difference in volts, V Q is the charge in coulombs, C

Fundamental physical quantities	
Physical quantity	**Unit(s)**
length	metre (m) kilometre (km) centimetre (cm) millimetre (mm)
mass	kilogram (kg) gram (g) milligram (mg)
time	second (s) millisecond (ms)
temperature	degree Celsius (°C) kelvin (K)
current	ampere (A) milliampere (mA)
voltage	volt (V) millivolt (mV)

Derived quantities and units	
Physical quantity	**Unit(s)**
area	cm^2; m^2
volume	cm^3; dm^3; m^3; litre (l); millilitre (ml)
density	kg/m^3; g/cm^3
force	newton (N)
speed	m/s; km/h
energy	joule (J); kilojoule (kJ); megajoule (MJ)
power	watt (W); kilowatt (kW); megawatt (MW)
frequency	hertz (Hz); kilohertz (kHz)
gravitational field strength	N/kg
radioactivity	becquerel (Bq)
acceleration	m/s^2; km/h^2
specific heat capacity	J/kg°C
specific latent heat	J/kg

Electrical symbols			
junction of conductors	ammeter	diode	capacitor
switch	voltmeter	electrolytic capacitor	relay (NO, COM, NC)
primary or secondary cell	indicator or light source	LDR	LED
battery of cells	or	thermistor	NOT gate (NOT)
power supply	motor	AND gate (AND)	OR gate (OR)
fuse	generator	NOR gate (NOR)	NAND gate (NAND)
fixed resistor	variable resistor		

Index